Observation and Modeling in Numerical Analysis and Model Tests in Dynamic Soil-Structure Interaction Problems

Proceedings of sessions
held in conjunction with Geo-Logan '97

W0082027

sponsored by The Geo-Institute of
the American Society of Civil Engineers

Logan, Utah
July 16-17, 1997

Edited by Toyoaki Nogami

Geotechnical Special Publication No. 64

GEO
INSTITUTE

Published by the

ASCE *American Society of Civil Engineers*

345 East 47th Street
New York, New York 10017-2398

Abstract:
Experimental and analytical studies have both involved soil-structure interaction. Experimental studies offer an opportunity to directly observe the complex phenomena, while analytical studies offer a logistic consideration on the physical insight into behavior and rational analysis methods. In these studies, however, there are many limitations as well as merits. In recent years, much progress has been made in the area concerning dynamic soil-structure interaction. The objective of this session was to discuss physical and numerical modeling in dealing with dynamic soil-structure interaction, in the light of recent developments. This proceedings of the session contains the papers on soil-structure interaction issues in centrifuge tests, shake table tests, full-scale tests, liquefaction and analytical aspects.

Library of Congress Cataloging-in-Publication Data

Observation and modeling in numerical analysis and model tests in dynamic soil-structure interaction problems : proceedings of sessions held in conjunction with Geo-Logan / sponsored by the Geo-Institute of the American Society of Civil Engineers, Logan, Utah, July 16-17, 1997 ; edited by Toyoaki Nogami.
p. cm. -- (Geotechnical special publication ; no. 64)
Includes indexes.
ISBN 0-7844-0252-3
1. Soil-structure interaction--Mathematical models--Congresses. 2. Soils--Vibration--Mathematical models--Congresses. I. Nogami, Toyoaki. II. American Society of Civil Engineers. Geo-Institute. III. Series.
TA711.5.027 1997   97-19515
624.1'5136--dc21   CIP

GEOTECHNICAL SPECIAL PUBLICATIONS

1) TERZAGHI LECTURES
2) GEOTECHNICAL ASPECTS OF STIFF AND HARD CLAYS
3) LANDSLIDE DAMS: PROCESSES, RISK, AND MITIGATION
4) TIEBACKS FOR BULKHEADS
5) SETTLEMENT OF SHALLOW FOUNDATION ON COHESIONLESS SOILS: DESIGN AND PERFORMANCE
6) USE OF IN SITU TESTS IN GEOTECHNICAL ENGINEERING
7) TIMBER BULKHEADS
8) FOUNDATIONS FOR TRANSMISSION LINE TOWERS
9) FOUNDATIONS AND EXCAVATIONS IN DECOMPOSED ROCK OF THE PIEDMONT PROVINCE
10) ENGINEERING ASPECTS OF SOIL EROSION, DISPERSIVE CLAYS AND LOESS
11) DYNAMIC RESPONSE OF PILE FOUNDATIONS— EXPERIMENT, ANALYSIS AND OBSERVATION
12) SOIL IMPROVEMENT - A TEN YEAR UPDATE
13) GEOTECHNICAL PRACTICE FOR SOLID WASTE DISPOSAL '87
14) GEOTECHNICAL ASPECTS OF KARST TERRAINS
15) MEASURED PERFORMANCE SHALLOW FOUNDATIONS
16) SPECIAL TOPICS IN FOUNDATIONS
17) SOIL PROPERTIES EVALUATION FROM CENTRIFUGAL MODELS
18) GEOSYNTHETICS FOR SOIL IMPROVEMENT
19) MINE INDUCED SUBSIDENCE: EFFECTS ON ENGINEERED STRUCTURES
20) EARTHQUAKE ENGINEERING & SOIL DYNAMICS (II)
21) HYDRAULIC FILL STRUCTURES
22) FOUNDATION ENGINEERING
23) PREDICTED AND OBSERVED AXIAL BEHAVIOR OF PILES
24) RESILIENT MODULI OF SOILS: LABORATORY CONDITIONS
25) DESIGN AND PERFORMANCE OF EARTH RETAINING STRUCTURES
26) WASTE CONTAINMENT SYSTEMS: CONSTRUCTION, REGULATION, AND PERFORMANCE
27) GEOTECHNICAL ENGINEERING CONGRESS
28) DETECTION OF AND CONSTRUCTION AT THE SOIL/ROCK INTERFACE
29) RECENT ADVANCES IN INSTRUMENTATION, DATA ACQUISITION AND TESTING IN SOIL DYNAMICS
30) GROUTING, SOIL IMPROVEMENT AND GEOSYNTHETICS
31) STABILITY AND PERFORMANCE OF SLOPES AND EMBANKMENTS II (A 25-YEAR PERSPECTIVE)
32) EMBANKMENT DAMS-JAMES L. SHERARD CONTRIBUTIONS
33) EXCAVATION AND SUPPORT FOR THE URBAN INFRASTRUCTURE
34) PILES UNDER DYNAMIC LOADS
35) GEOTECHNICAL PRACTICE IN DAM REHABILITATION
36) FLY ASH FOR SOIL IMPROVEMENT
37) ADVANCES IN SITE CHARACTERIZATION: DATA ACQUISITION, DATA MANAGEMENT AND DATA INTERPRETATION
38) DESIGN AND PERFORMANCE OF DEEP FOUNDATIONS: PILES AND PIERS IN SOIL AND SOFT ROCK

39) UNSATURATED SOILS
40) VERTICAL AND HORIZONTAL DEFORMATIONS OF FOUNDATIONS AND EMBANKMENTS
41) PREDICTED AND MEASURED BEHAVIOR OF FIVE SPREAD FOOTINGS ON SAND
42) SERVICEABILITY OF EARTH RETAINING STRUCTURES
43) FRACTURE MECHANICS APPLIED TO GEOTECHNICAL ENGINEERING
44) GROUND FAILURES UNDER SEISMIC CONDITIONS
45) IN-SITU DEEP SOIL IMPROVEMENT
46) GEOENVIRONMENT 2000
47) GEO-ENVIRONMENTAL ISSUES FACING THE AMERICAS
48) SOIL SUCTION APPLICATIONS IN GEOTECHNICAL ENGINEERING
49) SOIL IMPROVEMENT FOR EARTHQUAKE HAZARD MITIGATION
50) FOUNDATION UPGRADING AND REPAIR FOR INFRASTRUCTURE IMPROVEMENT
51) PERFORMANCE OF DEEP FOUNDATIONS UNDER SEISMIC LOADING
52) LANDSLIDES UNDER STATIC AND DYNAMIC CONDITIONS - ANALYSIS, MONITORING, AND MITIGATION
53) LANDFILL CLOSURES—ENVIRONMENTAL PROTECTION AND LAND RECOVERY
54) EARTHQUAKE DESIGN AND PERFORMANCE OF SOLID WASTE LANDFILLS
55) EARTHQUAKE-INDUCED MOVEMENTS AND SEISMIC REMEDIATION OF EXISTING FOUNDATIONS AND ABUTMENTS
56) STATIC AND DYNAMIC PROPERTIES OF GRAVELLY SOILS
57) VERIFICATION OF GEOTECHNICAL GROUTING
58) UNCERTAINTY IN THE GEOLOGIC ENVIRONMENT
59) ENGINEERED CONTAMINATED SOILS AND INTERACTION OF SOIL GEOMEMBRANES
60) ANALYSIS AND DESIGN OF RETAINING STRUCTURES AGAINST EARTHQUAKES
61) MEASURING AND MODELING TIME DEPENDENT SOIL BEHAVIOR
62) CASE HISTORIES OF GEOPHYSICS APPLIED TO CIVIL ENGINEERING AND PUBLIC POLICY
63) DESIGN WITH RESIDUAL MATERIALS; GEOTECHNICAL AND CONSTRUCTION CONSIDERATIONS
64) OBSERVATION AND MODELING IN NUMERICAL ANALYSIS AND MODEL TESTS IN DYNAMIC SOIL-STRUCTURE INTERACTION PROBLEMS
65) DREDGING AND MANAGEMENT OF DREDGED MATERIAL
66) GROUTING: COMPACTION, REMEDIATION AND TESTING
67) SPATIAL ANALYSIS IN SOIL DYNAMICS AND EARTHQUAKE ENGINEERING
68) UNSATURATED SOIL ENGINEERING PRACTICE
69) GROUND IMPROVEMENT, GROUND REINFORCEMENT, GROUND TREATMENT: Developments 1987-1997

Preface

Civil engineering structures are generally founded in or on the ground. The forces applied to, or induced in these structures by ground shaking are transmitted to the surrounding soil. Such soil-structure interaction often influences significantly the behavior of these structures. This interaction behavior is dependent on the structural properties, as well as the material properties and geometry of the ground. In a dynamic environment, soil-structure interaction generates out-going waves. Consequently, the wave propagation characteristics are also among the complex factors affecting behavior. In an earthquake ground shaking environment, the properties of free-field soil can change drastically, especially when the soil is liquefied. Hence, it is an extremely difficult task to properly take into account soil-structure interaction in structural response analysis and model tests simulating a dynamic environment.

There are experimental and analytical studies which involve soil-structure interaction. Experimental studies offer an opportunity to directly observe the complex phenomena, while analytical studies offer a logistic consideration on the physical insight into behavior and rational analysis methods. In these studies, however, there are many limitations as well as merits. In recent years, much progress has been made in the area concerning dynamic soil-structure interaction. The objective of the session is to discuss physical and numerical modeling in dealing with dynamic soil-structure interaction, in the light of recent developments.

It is the current practice of the Geo-Institute that each paper published in a Special Technical Publication (STP) be reviewed for its contents and quality. An STP is intended to reinforce the programs presented at convention sessions or specialty conferences and to contain papers that are timely and may be controversial to some extent. The time available for STP reviews is generally not as long and reviews may not be as comprehensive as those given to papers submitted to the Journal of Geotechnical and Environmental Engineering. These STP reviews ordinarily are carried out within a three month time frame. Therefore, it should be recognized that there is difference in the purpose and technical status of contributions to an STP, as compared to those in a journal. In accordance with ASCE policy, all papers published in this volume are eligible for discussion in the Journal of Geotechnical and Environmental Engineering and are eligible for ASCE awards. The following persons reviewed the papers published in this volume:

E. Aktan	S.E. Dickenson	P. Selvam
A. Anandarajah	A. Elgamar	R. Sen
R.M. Bakeer	M. Hynes	R. Siddharthan
A. Bodocsi	R.Y.S. Pak	C.V.G. Vallabhan
M. Budhu	S. Prakash	M. Vucetic

Thanks are extended to the authors of the papers for their biggest and most important job in the session. The editor would also like to express his gratitude to the many reviewers. Finally, thanks are due to Shiela Menaker who arranged for assembly and printing of this volume.

Toyoaki Nogami

Editor

Table of Contents

Multi-Layer Representation of Continuous Insitu Profiles in Soil Dynamics 1
B.B. Guzina and R.Y.S. Pak

Excess Pore Water Pressure Behind Quay Walls 11
Susumu Iai and Koji Ichii

Simplified Approach for Dynamic Soil-Structure Interaction Analysis of Rigid
Foundation 26
Toyoaki Nogami, Shengli Zhen, Atsushi Mikami and Kazuo Konagai

Aspects of Dynamic Centrifuge Testing of Soil-Pile-Superstructure Interaction 47
Daniel W. Wilson, Ross W. Boulanger, Bruce L. Kutter and Abbs Abghari

Axial Vibration of Circular Pile Foundations in Layered Soil Medium 64
C.V. Girija Vallabhan

Centrifuge and Numerical Modeling of Soil-Pile Interaction During Earthquake
Induced Soil Liquefaction and Lateral Spreading 76
T. Abdoun, R. Dobry and T.D. O'Rouke

Simulation of Soil-Structure Interaction on a Shaking Table 91
Kazuo Konagai and Toyoaki Nogami

Development of a Large-Scale Strong Motion Test Facility for Studying
Structural Response and Soil-Structure Interaction 107
*Pete Mote, Paul Gefken, Don Curran, Dave McCallen, Ignacio Arango
and Orhan Gurbuz*

Vibration Characteristics of Foundation Slabs on a Two-Parameter Elastic
Medium 138
*Musharraf Zaman, M. Omar Faruque, Adhir Agrawal, Weng Low
and Joakim G. Laguros*

Subject Index 139

Author Index 141

MULTI-LAYER REPRESENTATION OF CONTINUOUS INSITU PROFILES IN SOIL DYNAMICS

B. B. Guzina[1] and R. Y. S. Pak[1]

ABSTRACT

In the dynamic analysis of soil-structure interaction, an issue of common concern is the influence of insitu modulus variation on the foundation response. In three-dimensional problems, one of the questions that remain to be clarified is the effects of continuous wave refractions and curvilinear rays of travel in the case of a smoothly heterogeneous soil medium. This paper summarizes the results for the dynamic response of a footing on a soil with a linear shear velocity profile via (i) an exact treatment, and (ii) an approximate representation in terms of a piecewise homogeneous multi-layered system. Both solutions are generated by a boundary element method which treats the unbounded contact tractions along the foundation edges rigorously. The spatial variation of the soil's stiffness in each case is achieved through the use of exact Green's functions. The comparison of the numerical predictions supplies physical insights as well as modeling hints for the discretized approach to the soil dynamics problem. Specific aspects of the problem such as the occurrence of local resonance peaks and the phenomenon of cutoff frequency are also examined.

INTRODUCTION

In soil dynamics and earthquake engineering, it is well recognized that the soil profile can have a significant influence on the foundation and structural response under dynamic loading. While the soil's modulus profile can be continuous, discrete or a combination of both as a result of the deposition process and the gravity-induced stress conditions (Richart et al. 1971), the state-of-the-art analytical modeling of such insitu conditions is usually confined to a piecewise-homogeneous multi-layer representation. Due to a lack of more rigorous solution

---

[1]Department of Civil, Environmental and Architectural Engineering, University of Colorado, Boulder, CO 80309-0428, U.S.A.

for comparison, the accuracy of the solution and the degree of discretization needed for its achievement are often difficult to assess except on an ad-hoc basis. The purpose of this paper is to address such a problem and provide some critical insights on the validity and limitations of multi-layer approximations in soil dynamics.

In what follows, the basic formulation of a rigorous boundary element method is first introduced for the vibration analysis of a square surface footing (see Fig. 1). With the aid of exact Green's functions for a heterogeneous soil (Guzina and Pak 1996) and a multi-layer representation, a rigorous comparison is made on the similarity and differences between the two solutions.

METHOD OF ANALYSIS

For wave propagation problems involving semi-infinite media, boundary element formulations can be used as a firm foundation for the determination of both approximate and rigorous solutions for complicated boundary value problems. With the detailed derivation omitted for brevity, a frequency-domain version of the governing boundary integral equation can be given as

$$\int_\Gamma T_i(\boldsymbol{\xi}, \omega; \boldsymbol{n}) \, \hat{U}_i^k(\boldsymbol{\xi}, \boldsymbol{y}, \omega) \, d\Gamma_\xi \; - \; \int_\Gamma \left(U_i(\boldsymbol{\xi}, \omega) - U_i(\boldsymbol{y}, \omega) \right) [\hat{T}_i^k(\boldsymbol{\xi}, \boldsymbol{y}, \omega; \boldsymbol{n})]_1 \, d\Gamma_\xi$$

$$- \int_\Gamma U_i(\boldsymbol{\xi}, \omega) \, [\hat{T}_i^k(\boldsymbol{\xi}, \boldsymbol{y}, \omega; \boldsymbol{n})]_2 \, d\Gamma_\xi \; = \; U_k(\boldsymbol{y}, \omega). \tag{1}$$

In the above, ω is the circular frequency of vibration; $\boldsymbol{n}$ is the outward normal to the boundary surface Γ; $\hat{T}_i^k(\boldsymbol{\xi}, \boldsymbol{y}, \omega; \boldsymbol{n})$ and $\hat{U}_i^k(\boldsymbol{\xi}, \boldsymbol{y}, \omega)$ represent the appropriate traction and displacement Green's functions for the semi-infinite soil medium; and subscripts "1" and "2" denote their analytically-evaluated singular and numerically-evaluated regular parts, respectively. Upon the appropriate discretization of the boundary Γ, the above expression, which is free of Cauchy-principal values, can be solved numerically for the unknown boundary values of traction $T_i(\boldsymbol{\xi}, \omega; \boldsymbol{n})$ and displacement components $U_i(\boldsymbol{\xi}, \omega)$. Since the free-surface condition has been incorporated into the foregoing half-space fundamental solutions, only the soil-foundation interface needs to be discretized for the boundary integral formulation. To capture the unbounded contact stresses along the foundation edges and corners (see Fig. 5) which can contribute significantly to the foundation resistance, a set of special surface boundary elements whose traction shape functions are locally power-type singular have been designed and incorporated into the treatment. In the next section, numerical results and their implications pertaining to a square footing on a linear-wave-velocity half-space versus a multi-layered soil systems with 4, 8, and 64 layers are presented and discussed.

NUMERICAL RESULTS

The coordinate system and the geometry of a square rigid foundation with a dimension of $2a \times 2a$ are illustrated in Fig 1. The exact solution for a smoothly heterogeneous elastic half-space with a constant mass density ρ, shear modulus G, and a Poisson's ratio ν as

$$G(z) = G_0(1 + bz)^2, \qquad \nu = 0.25, \qquad (2)$$

will be used as a benchmark for comparison. In the approximate solution in which the soil is replaced by a piecewise homogeneous multi-layered system, the soil's modulus variation is modeled up to the depth of $H = 16a$. The material properties of the layers are chosen such that the shear modulus in each layer is equal to the exact value of (2) at the middle of the thickness (see Fig. 2). The bottom half-space, whose surface is located at the depth $z = H$, has a Poisson's ratio equal to 0.25 and a shear modulus $G_{hs} = G_0(1 + bH)^2$. The numerical solution for the piecewise homogeneous system is generated using the exact Green's functions for a multi-layered half-space (Guzina 1996).

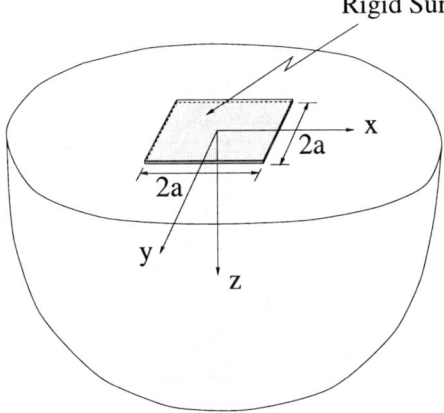

Figure 1: Geometry of Surface Foundation and Coordinate System

To represent the contact stress variations between the soil and the footing, the soil-foundation interface is modeled by a 15x15 mesh depicted in Fig. 3. The convergence of the approach is illustrated in Fig. 4 for the case of an exact treatment of a smoothly heterogeneous medium. Aided by the use of singular elements along the foundation edges, the results exhibit fast and stable convergence even for coarser meshes. In complex notation, a typical distribution of the dynamic contact stresses is illustrated in Fig. 5 from which the expected unbounded behavior of contact tractions can be readily observed.

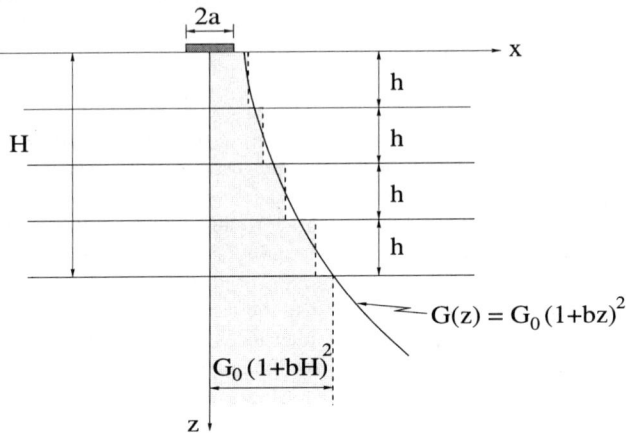

Figure 2: Multi-Layer Approximation of a Smoothly Heterogeneous Half-Space

In Figs. 6 to 10, the fundamental interfacial compliance functions (C_{vv}, C_{hh}, C_{mm}, C_{mh} and C_{tt}) pertaining to a square footing a the linear-wave-velocity half-space with $b = 0.10/a$ are presented. In the figures, the analytical benchmark solution for the smoothly heterogeneous soil is compared to approximations realized through a multi-layered soil representation with 4, 8, and 64 layers, respectively. For completeness, the results for a uniform half-space with $G = G_0$ and $\nu = 0.25$ are also included.

Considering the translational modes illustrated in Figs. 6 and 7, several observations are pertinent. As indicated by the display, the exact heterogeneous model and the 64-layer approximation show a high degree of agreement throughout the frequency range. In contrast to the homogeneous half-space solution whose real component decreases monotonically with frequency, the real parts of all heterogeneous models exhibit clear local maxima. Similar observations can be made for the magnitudes of the compliance functions C_{vv} and C_{hh}. Moreover, one can see additional oscillations in the dynamic compliance for a coarse multi-layer representation due to the resulting size of the material discontinuities. Another observation is that the exact solution indicates an existence of a cutoff frequency which is absent in the case of a uniform half-space. As can be expected from the depth-wise increase in the shear modulus in the benchmark solution, the homogeneous solution with $G = G_0$ underestimates the foundation stiffness at low frequencies, while the multi-layer model yields a

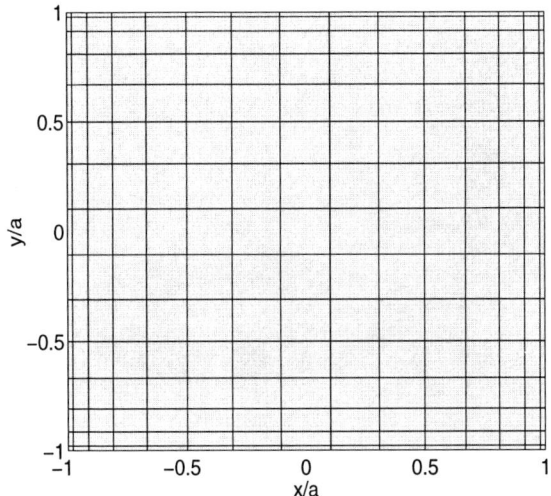

Figure 3: Discretization of the Interface Between the Soil and the Foundation

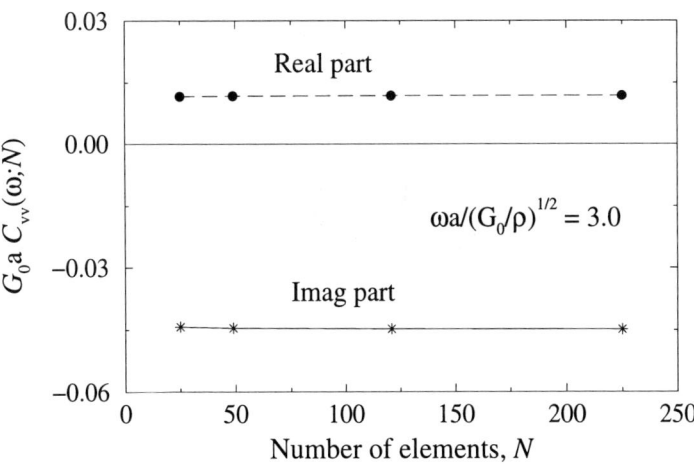

Figure 4: Dynamic Stiffness C_{vv} versus Number of Elements

Real Part

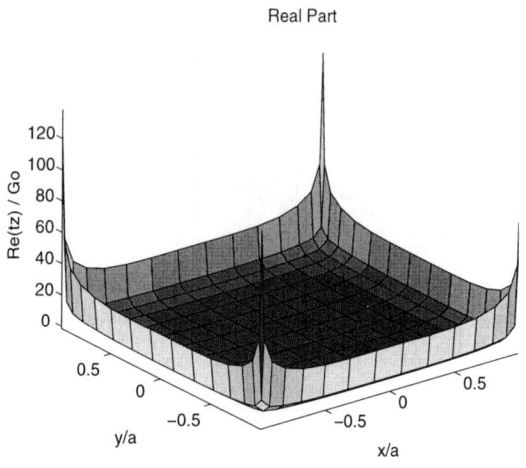

Imag Part

$\omega a/(G_0/\rho)^{1/2} = 3.0$

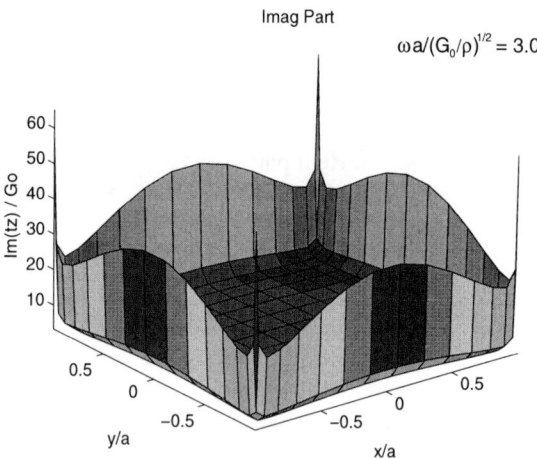

Figure 5: Distribution of Normal Contact Stress $t_z(x, y)$ per Unit Displacement Beneath the Rigid Foundation (Smoothly Heterogeneous Medium)

low-frequency behavior which is "stiffer" than that of the exact treatment. At high frequencies, however, the homogeneous half-space model appears to converge to the "exact" solution as it has the same surface modulus which is critical for small wave lengths. Because of the discretization scheme in Fig. 2, the 4- and 8-layer approximations consequently exhibit stiffer response than that of a smoothly heterogeneous solution even at high frequencies.

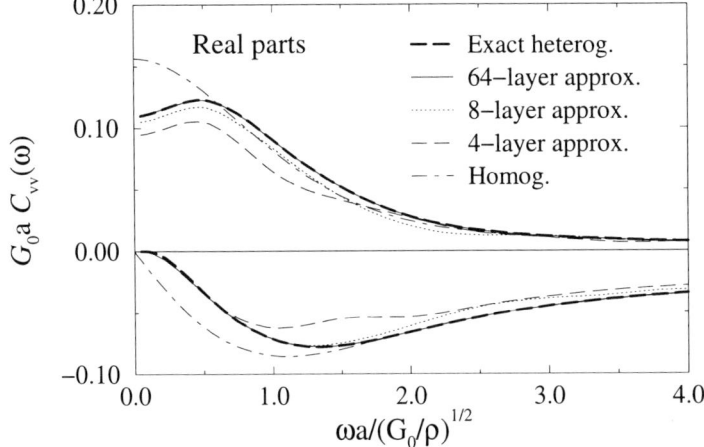

Figure 6: Vertical Compliance C_{vv}, Square Surface Foundation

In the case of the rocking compliance C_{mm} as depicted in Fig. 8, the exact heterogeneous model and the 64-layer approximation still compare well at all frequencies. However, the agreement between the exact and the coarser 4-layer and 8-layer approximations in this case is considerably worse than those for the translational modes in terms of the overall trend and general characteristics. This aptly indicates that for the common case where the foundation might rock, attention must be paid to capture the variation of the soil profiles in greater detail via insitu testing as well as in numerical modeling. As in the translational cases, the coarser multi-layer solutions would overestimate the low-frequency foundation stiffness. Due to the marked shift in the local maximum, however, these solutions yield significantly higher values of the interfacial compliance at higher frequencies.

The agreement between the "exact" solution and the 64-layer model persists for the remaining interfacial compliance functions C_{mh} and C_{tt} as shown in Figs. 9 and 10, respectively. By comparing the different models in Fig. 10, however, it is interesting to note that the torsional compliance function is strongly governed by the surface modulus throughout the frequency range of interest.

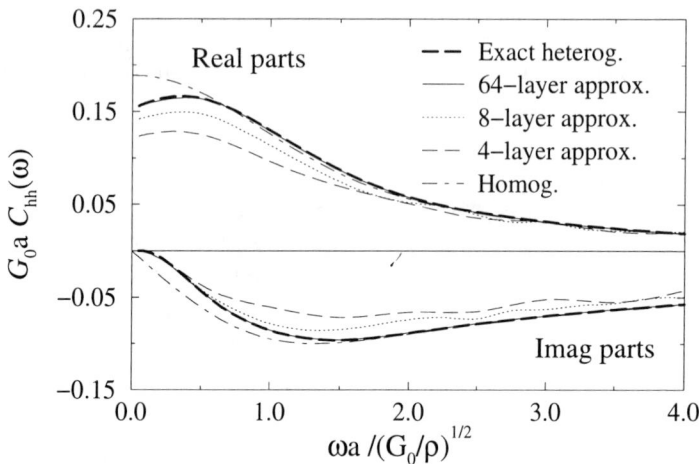

Figure 7: Lateral Compliance C_{hh}, Square Surface Foundation

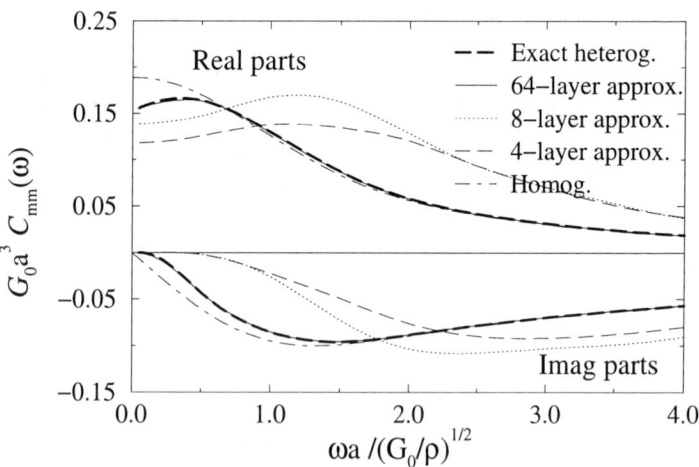

Figure 8: Rocking Compliance C_{mm}, Square Surface Foundation

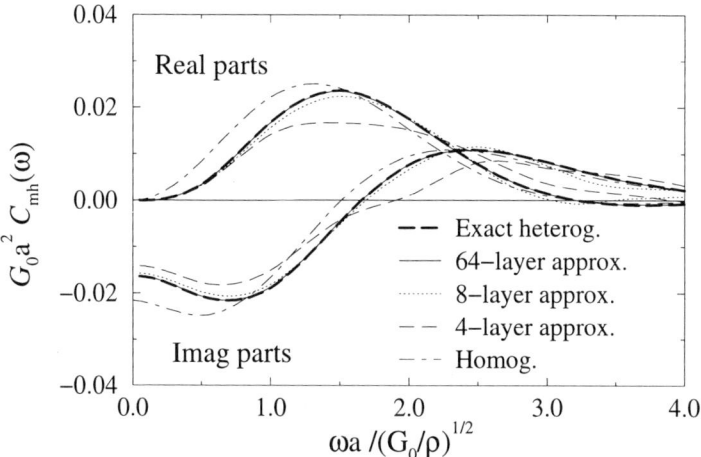

Figure 9: Coupling Compliance C_{mh}, Square Surface Foundation

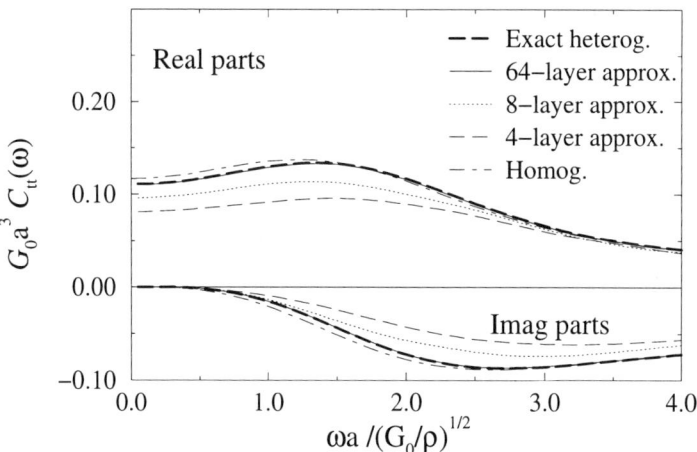

Figure 10: Torsional Compliance C_{tt}, Square Surface Foundation

SUMMARY AND CONCLUSIONS

In this paper, a new rigorous solution for a linear wave-velocity half-space is used as a basis to evaluate the accuracy of a piecewise-homogeneous multi-layered approximation in modeling the wave propagation in soils with a smooth continuous profile. By means of a direct boundary element method and appropriate set of Green's functions, the dynamic response of a square footing on a soil with a linear shear velocity profile is examined via (i) an exact treatment, and (ii) an approximate multi-layered model. Specific features of the foundation response pertaining to heterogeneous media such as the existence of the cutoff frequency and the local maxima of interfacial compliance functions are highlighted. The comparison of the numerical results indicates that a coarse multi-layered approximation can yield responses that are significantly different from the target, generally resulting in an underestimation of the foundation stiffness at low frequencies. A finely discretized multi-layer approximation to sufficient depth, on the other hand, does show excellent overall agreement with the results for a smoothly heterogeneous medium. It was also shown that the uniform half-space model whose modulus is equal to the surface modulus of a linear-wave-velocity half-space can be used as a good approximation at higher frequencies when most of the wave energy is radiated near the surface of the soil region. In general, however, accurate dynamic soil-structure interaction analysis can only be established with sufficient information on the insitu soil profile and an adequate representation of it in the numerical model.

REFERENCES

1. R. E. Richart, Hall and Wood (1971). *"Vibrations of Soils and Foundations"*, Prentice Hall.

2. B.B. Guzina and R.Y.S Pak (1996). *"Elastodynamic Green's Functions for a Smoothly Heterogeneous Half-Space"*, Int. J. Solids Struct., 33 (7), 1005-1021.

3. B.B. Guzina (1996). *"Seismic Response of Foundations and Structures in Multilayered Media."* Ph.D. Thesis, University of Colorado at Boulder, Boulder, Colorado, 365 pages.

EXCESS PORE WATER PRESSURES BEHIND QUAY WALLS

Susumu Iai[1] and Koji Ichii[2]

ABSTRACT

Excess pore water pressure increase in the loose saturated sand backfill behind quay walls can be quite different depending on the movement of the wall. If the wall movement is restricted by a rigidly fixed anchor or other structural means, the backfill sand can liquefy during strong shaking. If the wall is easier to move, however, the excess pore water pressures in the backfill sand may be absorbed by the seaward movement of the wall, never achieve the state of complete liquefaction. A caisson type quay wall founded on a loose liquefiable soil is an example. The effective stress analyses of two case histories during earthquakes in Japan, with and without restriction on the wall movement, are reviewed in this paper to discuss this subject. Results of the analyses demonstrate the capability of the effective stress model for understanding these complex soil-structure interaction phenomena.

INTRODUCTION

The seismic performance of retaining structures backfilled with dry sand is relatively well understood based on the Mohr-Coulomb's failure criterion and the Newmark's sliding block concept. The recent developments in this subject can be found in the state-of-the-art report by Whitman (1991). When the sand is saturated with water, as in waterfront structures, the effect of pore water pressure becomes predominant and complicates the performance of retaining structures. A simplified method, which combine the effect of pore water pressure and the sliding block concept may still be used if reasonable engineering judgements are used in determining the level of excess pore water pressure. Many laboratory data indicate, however, that the pore water pressure will not generally remain constant once deformation is induced in the sand. Dilatancy of sand complicates the behaviour of saturated sand under cyclic loading. Effective stress analysis may be the only reasonable means for taking the complicated behaviour of sand into account. In particular, a model consisting of a multiple shear mechanism (Iai et al,

[1]Chief & [2]Member, Geotechnical Earthquake Engineering Laboratory, Port and Harbour Research Institute, Ministry of Transport, Nagase 3-1-1, Yokosuka, 239 Japan

1992a and b) offers the prospect of taking into account the essential features of sand under cyclic loading.

Two case histories in Japan are reviewed in this paper to discuss the effect of the wall movement on the excess pore water pressure increase in the backfill sand. One is a sheet pile wall having backfill sand completely liquefied. The wall movement was restricted with a tie-rod and an anchor firmly installed in a non-liquefiable subsoil behind the backfill area. The other is a caisson wall having also a loose saturated backfill sand. Evidence of liquefaction, however, was not found in the backfill area within 30 m from the wall although abundant evidence of liquefaction was observed further inland. Effective stress analyses performed on these case histories are reviewed to discuss the capability of numerical modelling to represent these complicated soil-structure interaction phenomena of wall-soil systems.

QUAY WALLS SUPPORTED BY ANCHOR

In 1983, the Nihonkai Chubu Earthquake of magnitude 7.7 hit the northern part of Japan, caused serious damage to sheet pile quay walls at Akita Port located about 100 km from the epicentre (Iai and Kameoka, 1993). The maximum accelerations recorded at Akita Port were 219, 235, and 54 Gals in NS, EW, and UD directions. A typical cross section of the sheet pile quay walls at Akita Port is shown in Fig. 1. As shown in this figure, the quay wall was constructed with a backfilling method. The SPT N-values of the backfill sand, designated Layer 1 in Fig. 1, ranged from 2 to 10 whereas those of the original subsoil layer, designated Layer 2 in Fig. 1, underlying the backfill sand were greater than about 20.

Many sand boils were seen at the quay wall as shown in Fig. 2, suggesting liquefaction of backfill sand. The liquefaction caused serious damage to the sheet pile wall as shown by the broken lines in Fig. 1. Evidently an excessive large pressure was applied from the liquefied backfill to the sheet pile wall, resulting in yielding of the sheet pile. The top of the sheet pile wall was connected with a tie-rod to the anchor piles which were installed in the firm original ground beneath the backfill sand layer. Horizontal displacements at the top of the wall ranged from 1.1 to 1.8 m, evidently associated with those of the anchor piles; the anchors were pulled by the sheet piles towards the sea, accordingly inclined about 5 degrees presumably due to the reduced resistance in the liquefied backfill sand as shown in Fig. 1.

QUAY WALLS ON LOOSE SATURATED SAND

During the Great Hanshin earthquake of 1995, the caisson walls at Kobe Port were shaken with a strong earthquake motion having the peak ground accelerations of 0.54g and 0.45g in the horizontal and vertical directions. Most of the caisson walls were constructed on a loose saturated backfill sand, which was used for replacing the soft clayey deposit in Kobe Port in order to attain the required bearing capacity of foundation to support the caisson walls. These caisson walls displaced toward the sea about 5 m maximum, about 3 m average, settled about 1 to 2 m, and tilted about 4 degrees toward

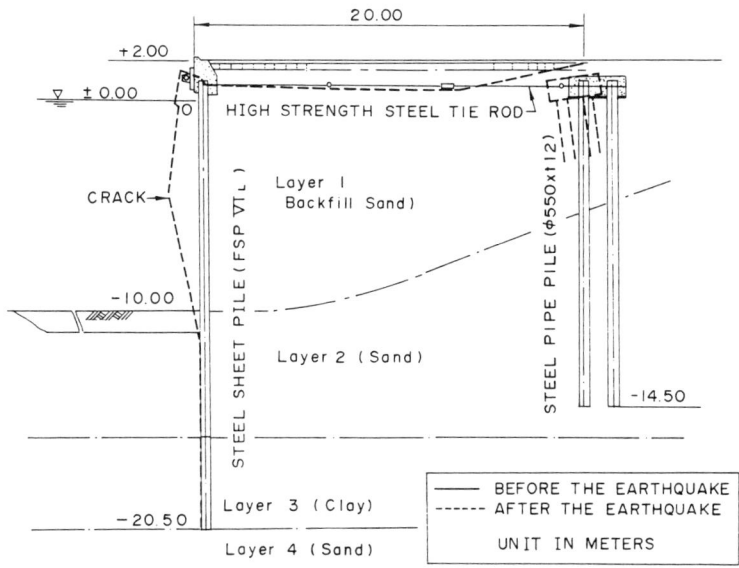

Fig. 1   Cross Section and Deformation of a Sheet Pile Quay Wall at Akita Port.

Fig. 2   Evidence on Liquefaction behind a Sheet Pile Quay Wall at Akita Port.

the sea. Figure 3 shows a typical cross section and deformation of the caisson walls at Kobe Port. There was little evidence of liquefaction at the backfill in the vicinity of the caisson walls as shown in Fig. 4 whereas extensive evidence of liquefaction of landfill soil was observed at inland about 30 m or further from the walls (Inagaki et al, 1996).

Although the sliding mechanism could explain the large horizontal displacement of the caisson walls, this mechanism did not explain the large settlement and tilting of the caissons. Reduction in the bearing capacity of foundation soils due to excess pore water pressure increase, then, was speculated as a main cause of the damage to the caisson walls. Excess pore water pressures in the backfill sand within 30 m from the caisson wall was presumed to be absorbed by the large seaward movement of the wall.

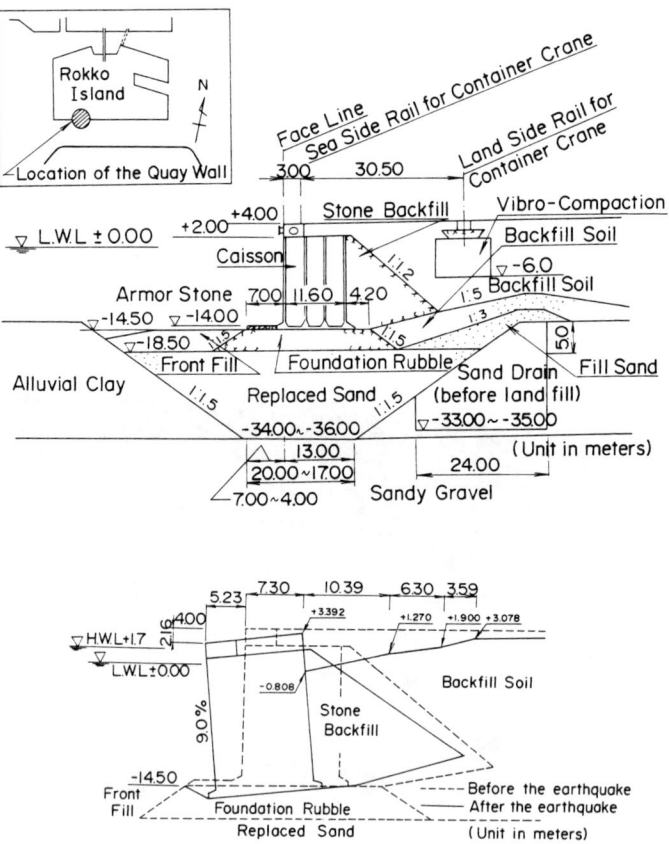

Fig. 3 Cross Section and Deformation of a Caisson Type Quay Wall at Kobe Port

MODELLING OF SOILS

In order to discuss the details of the above mentioned soil-structure interaction phenomena and evaluate the capability of numerical modelling for analyzing these phenomena, effective stress analyses were performed on these case histories. As mentioned earlier, effective stress model used was a multiple shear mechanism model defined in strain space (Iai et al, 1992a). This model has a capability to simulate the behaviour of sand subjected to rotation of principal stress axis, which plays an important role in the behaviour of initially anisotropically consolidated sand under cyclic simple shear (Iai et al, 1992b). With the effective stress and strain vectors written by

$$\{\sigma'\}^T = \{\sigma'_x \ \sigma'_y \ \tau_{xy}\} \tag{1}$$

$$\{\varepsilon\}^T = \{\varepsilon_x \ \varepsilon_y \ \varepsilon_{xy}\} \tag{2}$$

the basic form of the constitutive relation is given by

$$\{d\sigma'\} = [D](\{d\varepsilon\} - \{d\varepsilon_p\}) \tag{3}$$

in which

$$[D] = K\{n^{(0)}\}\{n^{(0)}\}^T + \sum_{i=1}^{I} R_{L/U}^{(i)}\{n^{(i)}\}\{n^{(i)}\}^T. \tag{4}$$

In this relation, the term $\{d\varepsilon_p\}$ in Eq. (3) represents the additional strain increment vector to take the dilatancy into account and is given from the volumetric strain increment due to the dilatancy $d\varepsilon_p$ as

$$\{d\varepsilon_p\}^T = \{d\varepsilon_p/2 \ d\varepsilon_p/2 \ 0\} \tag{5}$$

Fig. 4 Backfill Soil behind a Caisson Type Quay Wall at Kobe Port

The first term in Eq. (4) represents the volumetric mechanism with rebound modulus K and the direction vector is given by

$$\{n^{(0)}\}^T = \{1 \quad 1 \quad 0\}. \tag{6}$$

The second term in Eq. (4) represents the multiple shear mechanism. Each mechanism $i = 1,...,I$ represents a virtual simple shear mechanism, with each simple shear plane oriented at an angle $\theta_i/2 + \pi/4$ relative to the x axis. The tangential shear modulus $R^{(i)}_{L/U}$ represents the hyperbolic stress strain relationship with hysteresis characteristics. The direction vectors of the multiple shear mechanism in Eq. (4) are given by

$$\{n^{(i)}\}^T = \{\cos\theta_i \quad -\cos\theta_i \quad \sin\theta_i\} \quad \text{(for } i = 1,\cdots, I) \tag{7}$$

in which

$$\theta_i = (i-1)\Delta\theta \quad \text{(for } i = 1,\cdots, I) \tag{8}$$

$$\Delta\theta = \pi/I. \tag{9}$$

The loading and unloading for shear mechanism are separately defined for each mechanism by the sign of $\{n^{(i)}\}^T\{d\varepsilon\}$.

The parameters of the effective stress model, total of ten, were determined from the in-situ velocity logging, SPT N-values and the results of the cyclic triaxial tests using undisturbed samples. The constitutive model was incorporated into a finite element computer code and used for the numerical analysis. Before the earthquake response analysis, a static analysis was performed to simulate the stress conditions before the earthquake to take the effect of gravity into account. With the results of the static analyses used for initial conditions, earthquake response analyses were performed using the input earthquake motions recorded in the vicinity of the quay wall sites analyzed. Linear beam elements were used for the analyses of the sheet pile wall and the anchor piles for the sake of simplicity. For more details, see Iai and Kameoka (1993) for an analysis of sheet pile wall and Ichii et al. (1997) of a caisson wall.

As expected from the design practice of retaining walls, the initial stresses in the vicinity of the walls were very close to the condition of shear failure as shown later. Thus, the effective stress analyses in this study have to be able to capture the essential features of soil behaviour associated with the stress paths in the vicinity of the shear failure line. In order to review the capability of the effective stress model used in this study to analyze this type of behaviour, undrained cyclic simple shearing of initially anisotropically consolidated sand with $K_0 = 0.5$ was analyzed with keeping the initial axial stress difference unchanged for the stress boundary condition (Iai et al, 1992b).

The computed stress path in τ - $(-\sigma'_m)$ plane, shown in Fig. 5(a), indicates the gradual approach of effective stress path toward the failure line. In accordance with this approach, the stress path in τ_{xy} - $(-\sigma'_m)$ plane, shown in Fig. 5(b), clearly indicates a lower limit for $(-\sigma'_m)$. The stress and strain curve for τ_{xy} and γ_{xy}, shown in Fig. 5(c) indicates gradual growth in the amplitude of γ_{xy} whereas the axial strain difference (ε_x - ε_y), shown in Fig. 5(d), exhibits a cumulative increase. This cumulative increase of axial strain difference is a very important mechanism for governing the wall movement and settlements of soils behind the wall. All of these results for the sand under anisotropic stress conditions are consistent with the previous laboratory study by Ishihara and Li (1972), suggesting reasonable applicability of the present model for characterizing the cyclic behaviour of sand under anisotropic stress conditions close to the shear failure line.

ANALYSIS OF ANCHORED SHEET PILE WALL

The effective stress analysis of the anchored sheet pile quay wall at Akita Port resulted in the deformation shown in Fig. 6. Deformation towards the sea are mainly seen at the soils in front of and behind the sheet pile wall. As the shaking continues, the horizontal displacement at the top of the sheet pile gradually increased towards the sea and the backfill soil behind the sheet pile wall settled as shown in Fig. 7. The horizontal displacement at the top of the sheet pile wall was computed being about 1.3 m at the end of the shaking whereas the observed displacements range, as mentioned earlier, from

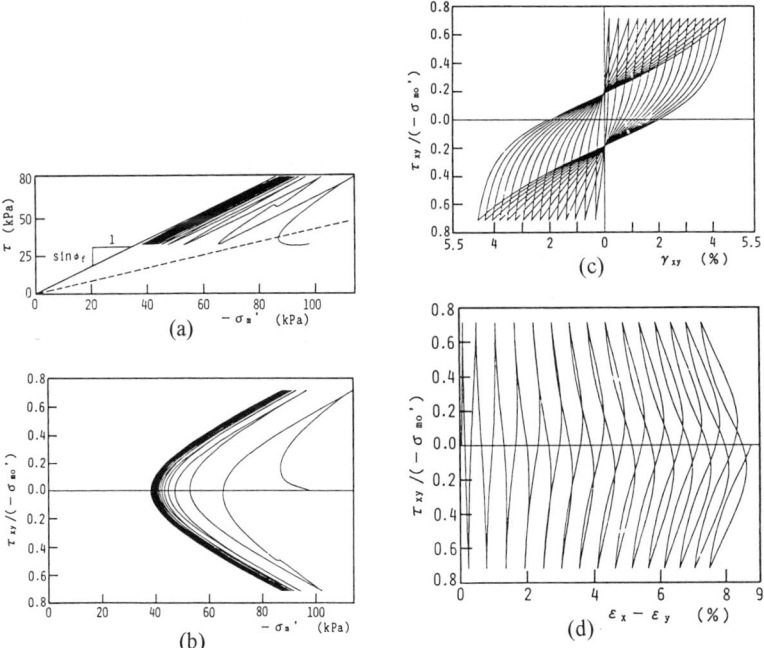

Fig. 5 Computed Stress Paths and Strains of a Sand Undergoing Cyclic Shearing under Anisotropic Stress Condition ($K_0 = 0.5$); (a) $\tau=(\sigma'_1-\sigma'_3)/2$ and $(-\sigma'_m)$, (b) $\tau_{xy}/(-\sigma'_{m0})$ and $(-\sigma'_m)$, (c) γ_{xy}, and (d) $(\varepsilon_x-\varepsilon_y)$

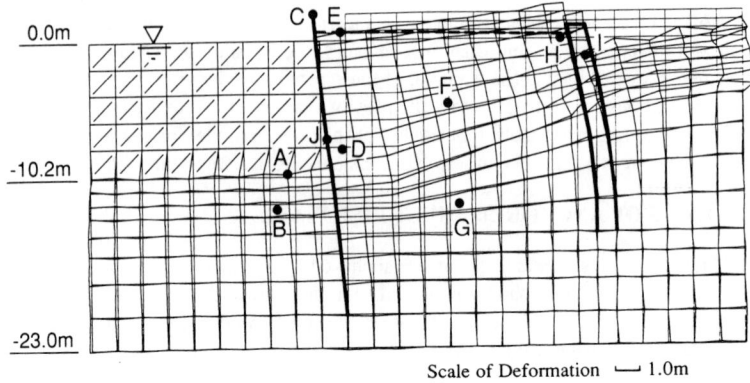

Fig. 6 Computed Residual Displacements of the Sheet Pile Quay Wall at Akita Port

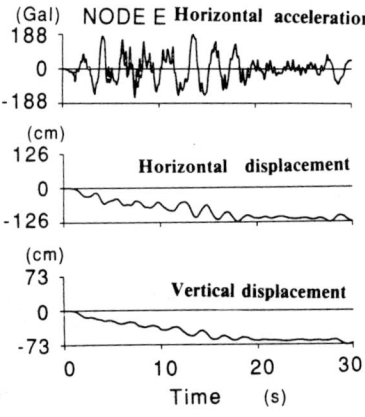

Fig. 7 Computed Response Acceleration and Displacements at the Backfill Soil just behind the Sheet Pile Wall (at Node E)

about 1.1 to 1.8 m, indicating the computed displacements are consistent with those observed.

As shown in Fig. 8, the earth pressures (i.e. the summations of the normal effective stresses and the excess pore water pressures) acting on the sheet pile wall became greater than those before the shaking. In particular the earth pressure behind the sheet pile after the shaking became about the same as those with the earth pressure coefficient of $K_0 = 1.0$. This is consistent with the results of the previous laboratory study on the earth pressures acting on the fixed wall due to the liquefied soil (Tsuchida, 1968). In accordance with those changes in the earth pressures, the bending moment of the sheet pile became greater as shown in Fig. 9. The computed maximum stress due to bending exceeded the yield strength of the steel sheet pile, indicating that the observed phenomenon, i.e. the yielding of the sheet pile, is well explained by the present analysis.

The excess water pressure in the soils in front of and behind the sheet pile wall were computed as shown in Fig. 10. The excess pore water pressures rapidly increased behind the sheet pile wall at the initiation of the shaking as shown in the second and third rows. The excess pore water pressures in front of the sheet pile wall, however, fluctuated around the value of zero as shown in the top row in Fig. 10.

In order to look into the mechanism of deformation and the excess pore water pressure increase of the quay wall, the stress and strain of soils in front of and behind the sheet pile wall, indicated by B and D in Fig. 6, are plotted in Fig. 11. The stress and strain notations are defined as extension being positive. In front of the sheet pile, the effective stress path, plotted as a relationship between the deviator stress $\tau = -(\sigma_1' - \sigma_3')/2$ and the effective mean stress $(-\sigma_m') = -(\sigma_x' + \sigma_y')/2$ shown in the upper most row in Fig. 11(a), gradually approaches the failure line with fluctuation around the initial deviator stress. Since the initial deviator stress was maintained close to the original level, reduction in the effective mean stress (i.e. excess pore water pressure increase) was barred by the shear failure line shown by a straight line with $1:\sin\phi_f$ slope in the same figure. In accordance with this stress path, the axial strain difference is gradually induced as shown in the middle row in the same figure. The shear strain γ_{xy} is also gradually induced but its magnitude is relatively small as shown in the bottom row in the same figure.

Behind the sheet pile wall, the effective mean stress $(-\sigma_m')$ rapidly decreases with a rapid release of initial deviator stress as shown in the upper most row in Fig. 11(b).

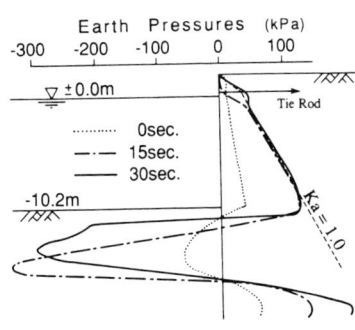

Fig. 8 Computed Earth Pressures on the Sheet Pile Wall

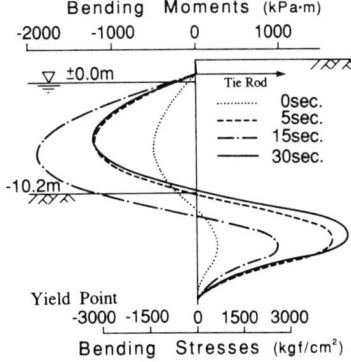

Fig. 9 Computed Bending Moments and Bending Stresses of the Sheet Pile Wall

Despite this rapid decrease of the initial deviator stress, the increase in the axial strain difference is slow and gradual as shown in the middle row in the same figure. Evidently, the rapid release in the deviator stress is due to the restricted movement of the wall, leading to complete liquefaction of soils retained by the wall.

ANALYSIS OF CAISSON WALL ON A LOOSE SAND

The effective stress analysis of a caisson wall constructed on a loose sand deposit shown in Fig. 3 resulted in the residual deformation shown in Fig. 12. As shown in this figure, the computed residual displacements of 3.5 and 1.5 m in horizontal and vertical directions with tilting of 4 degrees toward the sea were basically consistent with those observed and mentioned earlier. Computed mode of deformation of the caisson wall was to tilt into and push out the foundation soil beneath the caisson. This was also consistent with the observed deformation mode of the rubble foundation identified by a diving investigation (Inagaki et al, 1996).

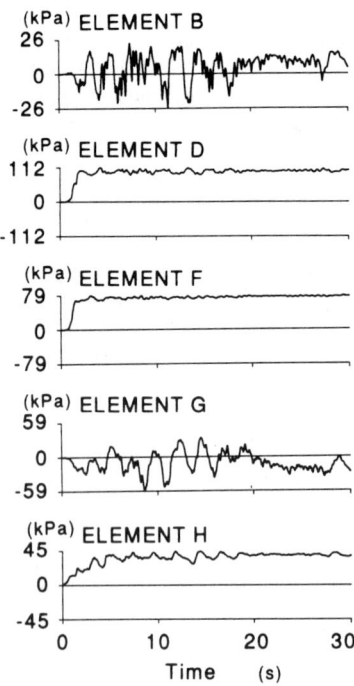

Fig. 10 Computed Excess Pore Water Pressures for the Sheet Pile Quay Wall

The computed displacements were gradually induced as shown in Fig. 13. The computed excess pore water pressures also gradually increased in the foundation soil as shown for Element A in Fig. 14 but never achieved the 100% level, in which the excess pore water pressure ratios were defined by $(1-\sigma'_m/\sigma'_{m0})$ using current and initial confining pressures σ'_m and σ'_{m0}. The definition of liquefaction has not been firmly established for the soil behaviours under anisotropic stress conditions. In this study, however, main concern is to try to explain the phenomenon of sand boils and water spring or the lack of them behind the walls. Thus, it may be sufficient here to use the excess pore water pressure ratios defined as above.

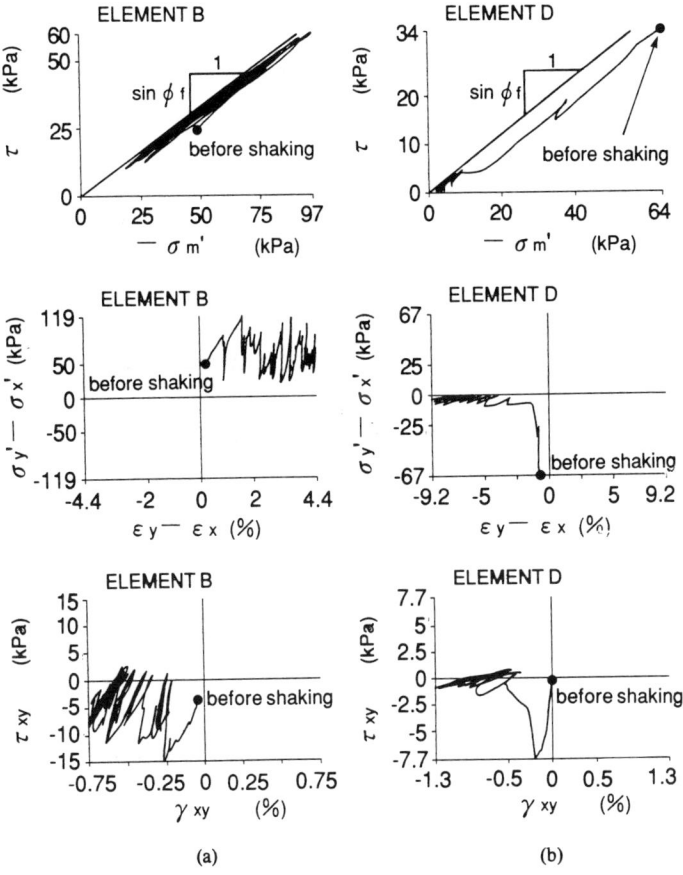

Fig. 11   Stress and Strain of Soils (a) in front of, and (b) behind, the Sheet Pile Wall

The computed excess pore water pressures just behind the caisson wall showed rapid increase but then decrease after about 8 seconds as shown for Element B in Fig. 14 and never achieved the 100% level whereas those at far inland area increased rapidly and achieved a level close to 100%. These computed excess pore water pressures were consistent with the observed performance of the backfill sand behind the quay wall mentioned earlier.

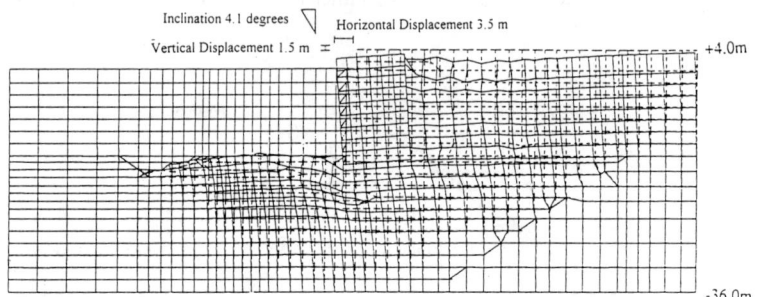

Fig. 12 Computed Residual Displacements of the Caisson Wall at Kobe Port

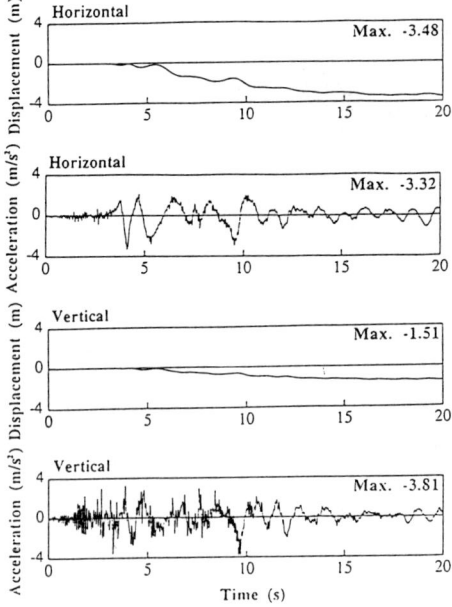

Fig. 13 Computed Response Accelerations and Displacements at the Top of the Caisson Wall

In order to discuss the mechanism of deformation and the excess pore water pressure increase of the quay walls, the stress and strain of soils below and behind the caisson wall are plotted in Fig. 15. Below the caisson wall (at Element A in Fig. 14(b)), the stress path shown in the upper most row in Fig. 15(a) gradually approaches the failure line with fluctuation around the initial deviator stress. Since the initial deviator stress was maintained close to the original level, reduction in the effective mean stress (i.e. excess pore water pressure increase) was barred by the shear failure condition. In accordance with this stress path, the axial strain difference is gradually induced as shown in the middle row in the same figure. The shear strain γ_{xy} is also gradually induced but its magnitude is relatively small as shown in the bottom row in the same figure.

Behind the caisson wall (at Element B in Fig. 14(b)), the effective mean stress $(-\sigma'_m)$ once decreased but then recovered close to the original level as shown in Fig. 15(b). This is presumably because the tensile strain in the horizontal direction was increased in the backfill sand due the seaward movement of the caisson wall, resulting in dilation of the sand and, thus the recovery of the effective mean stress. The normal strain difference governs the deformation of the backfill sand as shown in the same figure. Thus, when the wall is easier to move, the excess pore water pressures in the backfill soil behind the wall can be kept in to the low level, never to achieve the complete liquefaction.

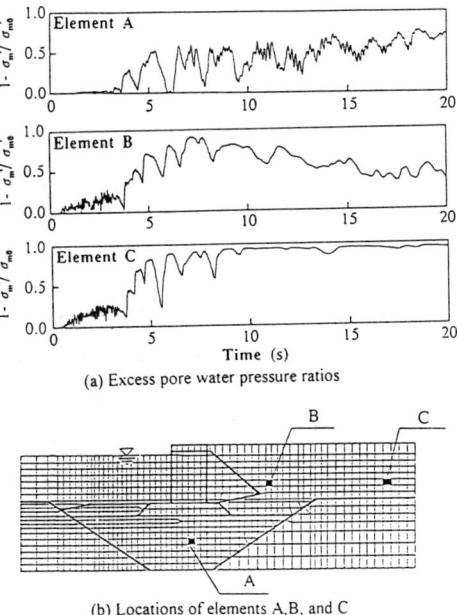

(a) Excess pore water pressure ratios

(b) Locations of elements A,B, and C

Fig. 14 Computed Excess Pore Water Pressure Ratios for the Caisson Type Quay Wall

As seen in the analyses of the two case histories of quay walls, the movement of the wall significantly affects the excess pore water pressures behind the wall. The effective stress analysis, which takes into account essential features of saturated soil-wall interaction phenomena, may be the only reasonable means to evaluated the excess pore water pressure levels behind the walls as well as the deformation of the soil-wall systems.

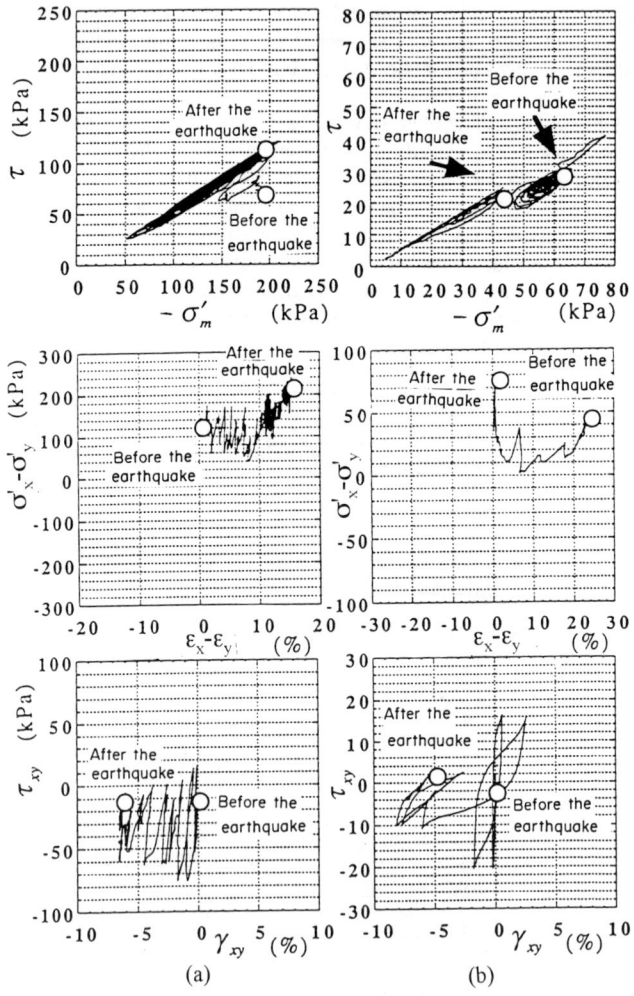

(a)   (b)

Fig. 15 Stress and Strain of Soils (a) beneath, and (b) just behind, the Caisson Wall

CONCLUSIONS

Effective stress analyses performed on two case histories in Japan to study the excess pore water pressure increase behind quay walls lead to the following conclusions.

1) When the wall movement is restricted by a rigidly fixed anchor or other means, loose saturated backfill sand behind the wall can liquefy during strong shaking. When the wall is easier to move, however, the excess pore water pressure increase in the backfill sand may be absorbed by the seaward movement of the wall, never to achieve the state of complete liquefaction.

2) Results of the effective stress analyses were consistent with the case histories of the quay walls in Japan, suggesting reasonable capability for numerical modelling of saturated soil-wall interaction phenomena, including evaluation of pore water pressure levels behind the walls and the deformation of the wall-soil systems.

REFERENCES

1. Iai, S., Matsunaga, Y. and Kameoka, T. (1992a). "Strain space plasticity model for cyclic mobility," Soils and Foundations, JSSMFE, 32(2), 1-15
2. Iai, S., Matsunaga, Y. and Kamokea, T. (1992b). "Analysis of undrained cyclic behaviour of sand under anisotropic consolidation," Soils and Foundations, JSSMFE, 32(2), 16-20
3. Iai, S. and Kameoka, T. (1993). "Finite element analysis of earthquake induced damage to anchored sheet pile quay walls," Soils and Foundations, JSSMFE, 33(1), 71-91
4. Inagaki, H., Iai, S., Sugano, T., Yamazaki, H., and Inatomi, T. (1996). "Performance of caisson type quay walls at Kobe Port," Soils and Foundations, Japanese Geotechnical Society, Special Issue, 119-136
5. Ichii, K., Iai, S. and Morita, T. (1997). "Analysis of deformation to gravity type quay walls using effective stress analysis," 2nd Conference on Great Hanshin-Awaji Earthquake of 1995, Japan Society of Civil Engineers, 251-258 (in Japanese)
6. Ishihara, K. and Li, S. "Liquefaction of saturated sand in triaxial torsion shear test," Soils and Foundations, JSSMFE, 12(2), 19-39
7. Tsuchida, H. (1968). "Earth pressure due to liquefied soil," Tsuchi-to-Kiso, JSSMFE, 16(5), 3-10 (in Japanese)
8. Whitman, R.V. (1991). "Seismic design of earth retaining structures," Proc. of the Second International Conference on Recent Advances in Geotechnical Earthquake Engineering and Soil Dynamics, St. Louis, 2, 1767-1778

Simplified Approach for
Dynamic Soil-Structure Interaction Analysis of Rigid Foundation

Toyoaki Nogami[1], Shengli Zhen[2], Atsushi Mikami[3] and Kazuo Konagai[4]

Abstract

Galerkin's procedure for weighted residual is applied to a simplified ground model. This results in a governing equation of the dynamic ground behavior only at the ground surface. The equation indicates that the ground surface behavior can be computed by an even further simplified model. By solving the governing equation for the boundary conditions along the surface, expressions in simple closed forms are developed for the dynamic response analysis of a massless rigid foundation which rests on the ground surface. Despite their significant simplicity, the developed expressions produce the computed values very close to those computed by far more complex rigorous solutions. They capture well the characteristics of the dynamic behavior of the ground.

Introduction

A large area of a foundation structure is generally in contact with soil to significantly modify the behavior of the foundation structure. Thus, it is important to take into account the ground behavior properly in the analysis of foundation response to external load. The dynamic loading applied to the foundation generates waves in the ground. As a result, the ground behavior is far more complex and a far larger area of the ground influences the foundation behavior in the dynamic condition than the static condition. Thus, it is extremely difficult to take into account the ground behavior in a simple, yet proper way in the analysis.

---

[1] Prof., Dept. of Civ. and Env. Engrg., Univ. of Cincinnati, Cincinnati, OH45221.

[2] Grad. Student, Dept. of Civ. and Env. Engrg., Univ. of Cincinnati, Cincinnati, OH45221.

[3] Research Assoc., Inst. of Industrial Science, Univ. of Tokyo, 7-22-1 Roppongi, Minato-Ku, Tokyo 106, Japan.

[4] Assoc. Prof., Inst. of Industrial Science, Univ. of Tokyo, 7-22-1 Roppongi, Minato-Ku, Tokyo 106, Japan.

The wave propagation effects in the ground can be directly consider if the ground is treated as a continuos medium in the analysis. With this treatment, wave equations were solved for the dynamic response behavior of ground by many people (e.g. Awojobi 1972; Karasudhi et al 1968; Luco 1976; Pak et al. 1991). This approach allows us to develop the solutions only for very simple ground conditions and the numerical evaluation of the solutions requires special mathematics which many practicing engineers are not familiar with.   The finite element method is a popular approach to compute the soil-structure interaction problems by treating the soil as a continuous medium (e.g. Lysmer et al. 1976).   A large area of the ground has to be discretized in FEM and thus, especially for the dynamic problem, it requires an extremely large computation effort.   A semi-analytical finite element method (Tassoulas 1985; Kausel and Roesset 1975) can reduce the computation effort to within a practical range but still requires a large computation effort which mostly results from solving the eigenvalue problem.   The boundary element method (BEM) was also used for the dynamic response analysis of foundations (Dominguez and Roesset 1978).   This method requires fundamental solutions, which are available only for very simple cases such as an infinite homogeneous medium.   Therefore, for realistic cases, BEM with typical fundamental solutions is still numerically intensive and not for routine usage.

The Winkler model is the simplest treatment for the ground in analyzing the dynamic responses of foundations and enables us to develop their formulations in very simple forms.   However, this approach fails to produce the complex characteristics of dynamic behavior of a continuous medium.   To overcome this deficiency, a special approximation was adopted for defining a Winkler model at the side of an embedded cylindrical foundation in which the ground medium was assumed to be made of uniformly distributed horizontal thin layers that were mutually uncoupled (Novak 1974). However, this model fails to produce the realistic behavior at frequencies below the fundamental frequency of the ground.   In order to improve this deficiency in the

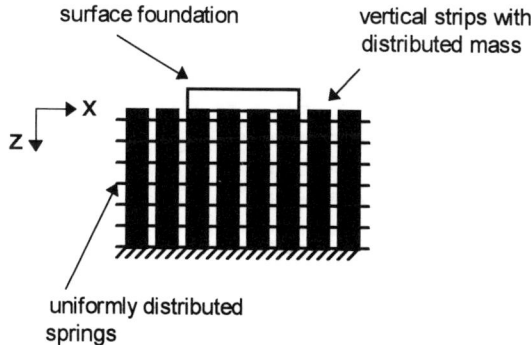

Fig. 1 Simplified Ground Model for
Dynamic Soil-Structure Interaction Analysis of Surface Foundation

dynamic response analysis of pile foundations, a coupling mechanism was introduced between the two adjacent springs in Winkler model (Nogami, et al. 1992). For surface foundations, a direct extension of Novak's treatment is to view the soil as a system of mutually uncoupled vertical strips which support the foundation on the top surface. A coupling mechanism was also introduced in between the two adjacent vertical strips in order to improve this model for surface foundations (Nogami and Leung 1990; Nogami 1996).

This paper presents simple formulations to compute the dynamic response of a two-dimensional massless rigid foundation on the ground surface. The formulations are based on the aforementioned ground model proposed by the first author.

Governing Equation for Dynamic Response of Ground Surface

(a) Simplified Model

Figure 1 shows schematically a previously proposed simplified ground model for the dynamic soil-structure interaction analysis (Nogami and Leung, 1990; Nogami 1996). The model is made of one-dimensional continuous vertical strips which are mutually interconnected by horizontal springs distributed along the vertical sides of the strips. When the vertical load is applied, normal forces are induced in the strips and the difference in displacements between the two adjacent strips produces shear spring forces. The displacements are only vertical displacements in this case. On the other hand, if the horizontal load is applied, horizontal shear forces are induced in the strips and the difference in displacements between the two adjacent strips produces normal spring forces. The displacements are only horizontal displacements in this case. The inertia force is induced by distributed mass in the vertical strips.

When the widths of the vertical strips are infinitesimally small, the strain in the strip and relative displacement between the two adjacent strips at (x, z) are expressed as respectively,

$$\frac{\partial u(x,z)}{\partial z} \quad \text{and} \quad \frac{\partial u(x,z)}{\partial x} \tag{1}$$

and thus the forces per unit area of the model, that associate with the deformations of the strips and springs at (x, z), are respectively

$$p_t(x,z) = k_t \frac{\partial u(x,z)}{\partial z} \quad \text{and} \quad p_s(x,z) = k_p \frac{\partial u(x,z)}{\partial x} \tag{2}$$

where k_t and k_p = stiffnesses per unit cross section area of strips and springs, respectively; p_t and p_p = force amplitudes; and u = vertical or horizontal displacement amplitude. In addition, the inertia force per unit volume of the strips at (x,z) is

$$p_m(x,z) = -\rho a(x,z) \qquad (3)$$

where a = acceleration amplitude (= $-\omega^2 u$); and ρ = mass per unit volume.

The parameters of the above model are stiffness (k_t) and mass (ρ) of the strip and stiffness (k_p) of the spring. It has been found that these stiffnesses are independent of both frequency and ground displacement shape but dependent on Poisson's ratio, and that mass is equal to that of the soil (Nogami and Leung 1990). Because the model parameters are independent of frequency and ground displacement shape, the stiffness parameters can be conveniently determined from static plate loading tests in the field. If the elastic moduli of the ground are known, they can be determined from theses moduli by the following expressions also (Nogami and Leung 1990):

$$k_p = C_p^* \overline{k}_p \qquad (4a)$$
$$k_t = C_t^* \overline{k}_t \qquad (4b)$$

where

vertical response

$$C_p^* = G(1+2Di) \qquad (5a)$$
$$C_t^* = (\lambda + 2G)(1+2Di) \qquad (5b)$$
$$\overline{k}_t = 0.96 - 0.49v + 6.47v^2 - 15.76v^3 \qquad (5c)$$
$$\overline{k}_p = 0.5\overline{k}_t \qquad (5d)$$

horizontal response

$$C_p^* = (\lambda + 2G)(1+2Di) \qquad (6a)$$
$$C_t^* = G(1+2Di) \qquad (6b)$$
$$\overline{k}_t = 0.85 - 0.50v + 4.22v^2 - 10.72v^3 \qquad (6c)$$
$$\overline{k}_p = \overline{k}_t \qquad (6d)$$

with G (shear modulus) and λ = Lame's constants of soil; and D and v = material damping factor and Poison's ratio of soil, respectively. It is noted that: k_t and k_p with $\overline{k}_t = 1$ and $\overline{k}_p = 1$ in the vertical response are the stiffnesses of a visco-elastic medium in normal and shear stress-strain relationships, respectively; and those in horizontal response are the stiffnesses of a visco-elastic medium in shear and normal stress-strain relationships, respectively.

Consideration of the equilibrium condition of these forces acting on a segment with small thichness, dz, in a single strip results in the following equation of motions of the model for the steady-state harmonic vibration:

$$k_t \frac{\partial^2 u(x,z)}{\partial z^2} + k_p \frac{\partial^2 u(x,z)}{\partial x^2} - \rho a(x,z) = 0 \tag{7}$$

(b) Development of Equation at the Surface

The displacement within the model is approximated by

$$u(x,z) = u(x)\phi(z) \tag{8}$$

where $\phi(z=0) = 1$; and $u(x)$ = ground surface displacement. Using the above function $\phi(z)$, Galerkin's procedure for weighted residual is applied to Eq. 7. This results in

$$\int_0^H \left(k_t u(x) \frac{d^2\phi(z)}{dz^2} + k_p \frac{d^2 u(x)}{dx^2} \phi(z) + \rho\omega^2 u(x)\phi(z) \right) \phi(z) dz = 0 \tag{9}$$

Green's second theorem leads Eq. 9 to

$$\int_0^H \left(-k_t u(x) \frac{d\phi(z)}{dz} \frac{d\phi(z)}{dz} + k_p \frac{d^2 u(x)}{dx^2} \phi^2(z) + \rho\omega^2 u(x)\phi^2(z) \right) dz +$$

$$k_t \frac{\partial u(x,z)}{\partial z} \phi(z) \Big|_{z=0}^H = 0 \tag{10}$$

In the above expression, $k_t\, \partial u/\partial z$ at $z = 0$ is equal to the external load acting on the surface, and $\phi(z)$ is equal to one and zero respectively at $z = 0$ and H. Thus, Eq. 10 is rewritten as

$$-N \frac{d^2 u(x)}{dx^2} + Ku(x) - \omega^2 Mu(x) = p(x) \tag{11}$$

where $p(x)$ = distributed load at the surface; and

$$N = k_p \int_0^H \phi^2(z) dz \tag{12a}$$

$$K = k_t \int_0^H (\phi'(z))^2 dz \tag{12b}$$

$$M = \rho \int_0^H \phi^2(z) dz \tag{12c}$$

with $\phi'(z) = d\phi(z)/dz$. Eq. 11 is a governing equation of the ground surface behavior which is caused by the dynamic load, $p(x)$, applied on the ground surface.

If the ground surface is free at x, $p(x)$ is equal to zero. If an Euler beam (flexible foundation) exists at x on the surface, $p(x)$ is

$$p(x) = -\left(EI \frac{d^4 u(x)}{dx^4} - \omega^2 m_b u(x) \right) + P(x) \qquad (13)$$

where $P(x)$ = amplitude of load applied to the beam; and m_b = mass of the beam per unit length. Considering the compatibility and equilibrium conditions along the soil-foundation interface, Eqs. 11 and 13 are combined to form a governing equation for the soil-foundation coupled behavior such that

$$EI \frac{d^4 u(x)}{dx^4} - N \frac{d^2 u(x)}{dx^2} + Ku(x) - \omega^2 (m_b + M)u(x) = P(x) \qquad (14)$$

Eq. 11 is solved to obtain simple expressions for soil stiffnesses associated with rigid foundations, while both Eqs. 11 and 14 are solved for flexible foundations on the ground surface. The function $\phi(z)$ is yet to be defined for solving these equations.

Function $\phi(z)$

Using the surface displacement, $u(x)$, defined in Eq. 8 as a function, Galerkin's procedure for weighted residual is applied to Eq. 7. This results in

$$\int_{-\infty}^{\infty} \left(k_t u(x) \frac{d^2 \phi(z)}{dz^2} + k_p \frac{d^2 u(x)}{dx^2} \phi(z) + \rho \omega^2 u(x)\phi(z) \right) u(x) dx = 0 \qquad (15)$$

Green's second theorem leads Eq. 6 to

$$\int_{-\infty}^{\infty} \left(k_t u^2(x) \frac{d^2 \phi(z)}{dz^2} - k_p \frac{du(x)}{dx} \frac{du(x)}{dx} \phi(z) + \rho \omega^2 u^2(x)\phi(z) \right) dx +$$

$$k_p \frac{\partial u(x,z)}{\partial x} \phi(z) \bigg|_{x=-\infty}^{\infty} = 0 \qquad (16)$$

In the above expression, $k_p \, \partial u/\partial x$ is equal to the spring force and is zero at x = $\pm\infty$ for loading over a limited area Thus, Eq. 16 is rewritten as

$$-n\frac{d^2\phi(z)}{dz^2} + k\phi(z) - \omega^2 m\phi(z) = 0 \qquad (17)$$

where, with $u'(x) = du(x)/dx$

$$n = k_s \int_{-\infty}^{\infty} u^2(x)dx \qquad (18a)$$

$$k = k_p \int_{-\infty}^{\infty} (u'(x))^2 dx \qquad (18b)$$

$$m = \rho \int_{-\infty}^{\infty} u^2(x)dx \qquad (18c)$$

The general solution of Eq. 17 for $\phi(z)$ is

$$\phi(z) = C_1 e^{-\alpha z} + C_2 e^{\alpha z} \qquad (19)$$

where C_1 and C_2 = unknown constants to be determined from the boundary conditions; and

$$\alpha^2 = \frac{k - \omega^2 m}{n} \qquad (20)$$

The boundary conditions for $\phi(z)$ are $\phi(z=0) = 1$ and $\phi(z=H) = 0$. After determining C_1 and C_2 from these conditions, the function $\phi(z)$ is defined as

$$\phi(z) = \frac{1}{e^{-2\alpha H} - 1}\left(e^{-2\alpha H} e^{\alpha z} - e^{-\alpha z}\right) \qquad (21)$$

It is noted that α is a function of the displacement along the ground surface, $u(x)$, and frequency, ω, according to Eqs. 18 and 20. The function, $\phi(z)$, is therefore dependent on frequency and the ground surface displacement.

Simple Expressions for Dynamic Response of a Massless Rigid Foundation

Eq. 11 indicates that, for the load applied on the ground surface, the ground behavior along the surface is produced by a further simplified model like the one shown in Fig. 2. This further simplified model is made of a one-dimensional horizontal beam supported by a vertical Winkler spring model. The model is characterized by the stiffness (N) and mass (M) of the beam and the stiffness of the spring (K). The deformation of the beam is a shear deformation for the vertical load and an axial deformation for the lateral load. Contrary to the parameters of the model shown in

Fig. 1, the parameters of this further simplified model are dependent on ω and u(x) as stated earlier.

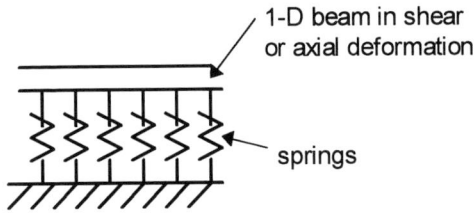

Fig. 2 Further Simplified Model

(a) Expressions

Simple expressions are developed for the dynamic responses of a massless rigid foundation on the ground surface, by solving Eq. 11 with the function $\phi(z)$ and the boundary conditions along the surface. The origin of the x and z coordinates is set at the center of the foundation base. The boundary conditions along the surface are

translational responses:

$$\begin{cases} p(x) = 0 & x \leq -b \text{ and } x \geq b \\ \int_{-b}^{b} p(x)dx = P_u \quad \text{and} \quad u(x) = u & -b \leq x \leq b \end{cases} \tag{22a}$$

rocking response:

$$\begin{cases} p(x) = 0 & x \leq -b \text{ and } x \geq d \\ \int_{-b}^{b} xp(x)dx = P_\theta \quad \text{and} \quad u(x) = x\theta & -b \leq x \leq b \end{cases} \tag{22b}$$

where b = half of the width of foundation; and u and θ = displacement amplitudes in translation and rotation of a rigid foundation, respectively; and P_u and P_θ = amplitudes of translational force and moment, respectively.

Eq. 11, with the boundary conditions stated in Eqs. 22a and 22b, is solved to obtain simple expressions for dynamic foundation responses (Appendix I). The finally obtained expressions in nondimensional forms are:

vertical response

$$\frac{P_u}{uG} = 2(1+2Di)\overline{k}_p(\overline{\beta N} + \overline{\kappa K}) - 2a_0^2\overline{M} \tag{23a}$$

$$\overline{\beta}^2 = \kappa\frac{-\overline{K}}{\overline{N}} - \frac{a_0^2}{1+2Di} \tag{23b}$$

$$\overline{\alpha}^2 = \frac{1}{\kappa}\left(\frac{\overline{k}}{\overline{n}} - \frac{a_0^2}{1+2Di}\right) = \frac{1}{\kappa}\left(\frac{\overline{\beta}^2}{1+2\overline{\beta}} - \frac{a_0^2}{1+2Di}\right) \tag{23c}$$

horizontal response

$$\frac{P_u}{uG} = 2(1+2Di)\overline{k}_t\left(\frac{\overline{\beta}}{\kappa}\overline{N} + \overline{K}\right) - 2a_0^2\overline{M} \tag{24a}$$

$$\overline{\beta}^2 = \kappa\left(\frac{\overline{K}}{\overline{N}} - \frac{a_0^2}{1+2Di}\right) \tag{24b}$$

$$\overline{\alpha}^2 = \frac{1}{\kappa}\frac{\overline{k}}{\overline{n}} - \frac{a_0^2}{1+2Di} = \frac{1}{\kappa}\frac{\overline{\beta}^2}{1+2\overline{\beta}} - \frac{a_0^2}{1+2Di} \tag{24c}$$

rocking response

$$\frac{P_\theta}{\theta Gb^2} = 2(1+2Di)\overline{k}_p\left(\overline{\beta N} + \frac{1}{3}\overline{kK}\right) - \frac{2}{3}a_0^2\overline{M} \tag{25a}$$

$$\overline{\beta}^2 = \kappa\frac{-\overline{K}}{\overline{N}} - \frac{a_0^2}{1+2Di} \tag{25b}$$

$$\overline{\alpha}^2 = \frac{1}{\kappa}\left(\frac{\overline{k}}{\overline{n}} - \frac{a_0^2}{1+2Di}\right) = \frac{1}{\kappa}\left(\frac{3\overline{\beta}^2 + 6\overline{\beta}}{3+2\overline{\beta}} - \frac{a_0^2}{1+2Di}\right) \tag{25c}$$

where $\overline{\kappa} = k_t / k_p$; $\overline{\beta} = \beta b$ with β defined in Eq. 30 in Appendix I; $\overline{\alpha} = \alpha b$; $\overline{n}$ and $\overline{k} =$ nondimensional forms of integrations in n and k (Eq. 18), respectively; and $a_0 = b\omega/v_s$ with $v_s =$ shear wave velocity of soil. $\overline{N}$, $\overline{K}$ and $\overline{M}$ are nondimensional forms of integrations in N, K and M (Eq. 12), respectively. With $\phi(z)$ in Eq. 21, they are expressed as

$$\overline{N} = \overline{M} = \int_0^{\overline{H}} \phi^2(z)d\overline{z} = -\left(\frac{1}{e^{-2\overline{\alpha H}} - 1}\right)^2\left\{2\overline{H}e^{-2\overline{\alpha H}} + \left(\frac{e^{-4\overline{\alpha H}}}{2\overline{\alpha}} - \frac{1}{2\overline{\alpha}}\right)\right\} \tag{26a}$$

$$\overline{K} = \int_0^{\overline{H}} \left(\frac{d\phi(\overline{z})}{d\overline{z}} \right)^2 d\overline{z} = \left(\frac{\overline{\alpha}}{e^{-2\overline{\alpha}\overline{H}} - 1} \right)^2 \left\{ 2\overline{H}e^{-2\overline{\alpha}\overline{H}} - \left(\frac{e^{-4\overline{\alpha}\overline{H}}}{2\overline{\alpha}} - \frac{1}{2\overline{\alpha}} \right) \right\} \tag{26b}$$

where $\overline{z} = z/b$; $\overline{N} = bN/k_p$; $\overline{K} = K/(bk_t)$; and $\overline{M} = bN/\rho$.

$\overline{\beta}$ and $\overline{\alpha}$ are coupled with each other in Eqs. 23b and 23c for vertical response, Eqs. 24b and 24c for horizontal response and Eqs. 25b and 25c for rocking response. In the computation, they are first determined from these coupled equations by iteration with the assumed initial value of either $\overline{\beta}$ or $\overline{\alpha}$. Then, the responses of a massless foundation are computed by Eq. 23a, Eq. 24a and Eq. 25a with these determined $\overline{\beta}$ and $\overline{\alpha}$. When the foundation contains a mass, the inertia force normalized in a similar manner is added on the right hand side of these equations.

For a half-space ground, substituting $\overline{H} = \infty$ into Eqs 26a and 26b results in

$$\overline{N} = \frac{1}{2\overline{\alpha}} \tag{27a}$$

$$\overline{K} = \frac{\overline{\alpha}}{2} \tag{27b}$$

and thus Eqs. 23b, 24b and 25b are rewritten as

$$\overline{\beta}^2 = \begin{cases} \overline{\kappa}\overline{\alpha}^2 - \dfrac{a_0^2}{1+2Di} & \text{vertical and rocking responses} \\[4mm] \overline{\kappa}\left(\overline{\alpha}^2 - \dfrac{a_0^2}{1+2Di} \right) & \text{horizontal response} \end{cases} \tag{27c}$$

In addition to Eq. 27c, Eqs. 23c, 24c and 25c express the relationship between $\overline{\alpha}$ and $\overline{\beta}$ also. Substituting these equations into Eq. 27c, the following equations are obtained:

$$\overline{\beta}^3 + \frac{2a_0^2}{1+2Di}\overline{\beta} + \frac{a_0^2}{1+2Di} = 0 \qquad \text{vertical response} \tag{28a}$$

$$\overline{\beta}^3 + \frac{1}{\overline{\kappa}}\frac{2a_0^2}{1+2Di}\overline{\beta} + \frac{1}{\overline{\kappa}}\frac{a_0^2}{1+2Di} = 0 \qquad \text{horizontal response} \tag{28b}$$

$$\overline{\beta}^3 - 3\overline{\beta} + \left(3+2\overline{\beta}\right)\frac{a_0^2}{1+2Di} = 0 \qquad \text{rocking response} \tag{28c}$$

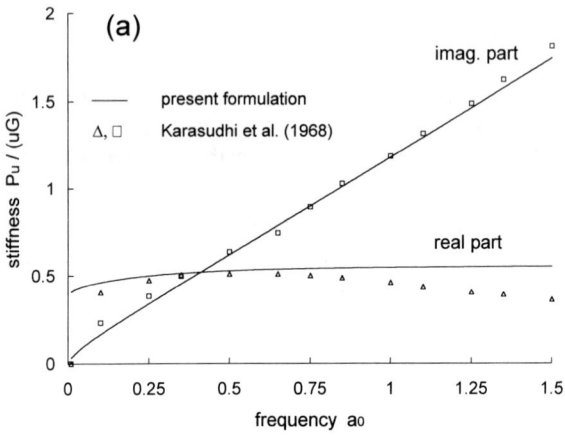

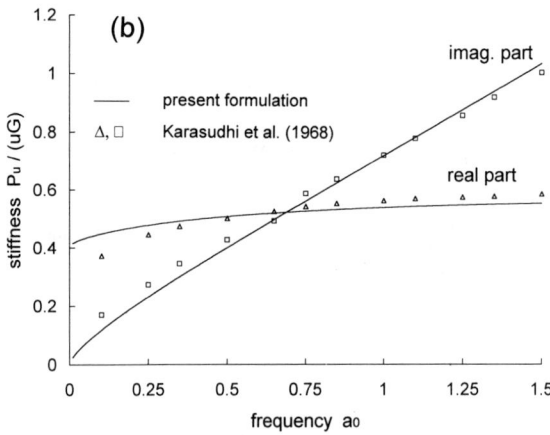

Fig. 3 Impedance Functions for Foundation on Half Space:
(a)Vertical Response; (b) Horizontal Response

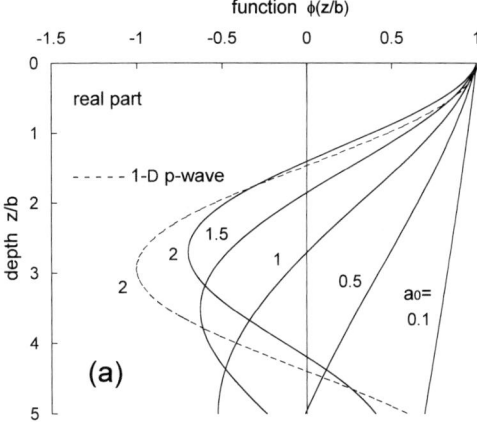

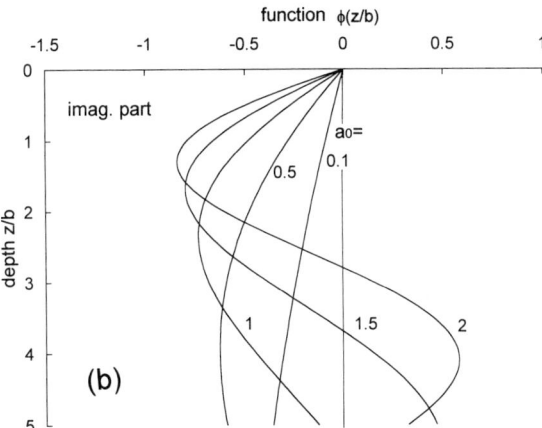

Fig. 4 Function $\phi(z)$ for Vertical Response of Foundation on Half-Space
at Various Frequencies: (a) Real Part; (b) Imaginary Part

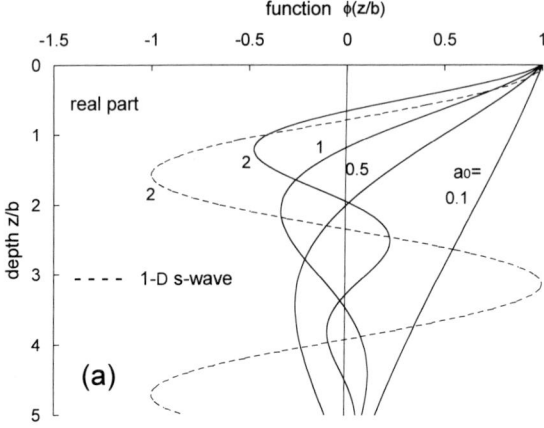

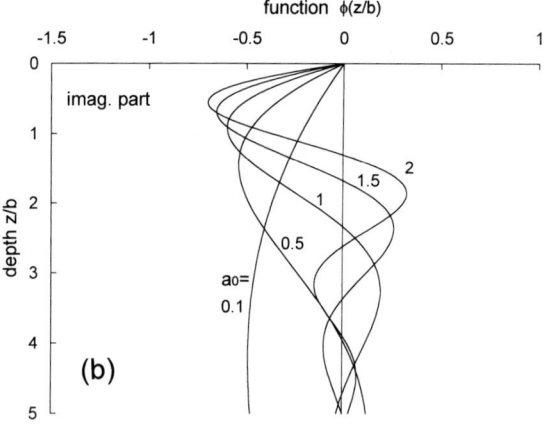

Fig. 5 Function $\phi(z)$ for Horizontal Response of Foundation on Half-Space
at Various Frequencies: (a) Real Part; (b) Imaginary Part

Therefore, $\overline{\beta}$ for a half-space ground is determined directly from Eqs. 28a ~ 28c without iteration. Then, $\overline{\alpha}$ is determined from Eqs. 24c, 25c and 26c with the defined $\overline{\beta}$.

(b) Computed Results and Remarks

    Dynamic responses of a massless rigid strip foundation on an elastic half-space are computed by using the above obtained simple expressions. $D = 2\%$ and $\nu = 1/3$ are assumed for soil. Fig. 3 shows the computed impedance functions (stiffness ~ a_0 relationship) of this foundation for vertical and horizontal responses. The results computed by a more rigorous approach (Karasudhi et al. 1968) are also shown by dots. Despite significant simplification, the present approach produces the results very close to those computed by rigorous solutions. Figs. 4 and 5 show the function, $\phi(z)$, at various frequencies for vertical and horizontal responses, respectively. As seen in the figure, the function depends on frequency. Broken lines in Fig. 4 and 5 show respectively the variations of P-wave and S-wave displacements along the depth at $a_0 = 2.0$ when these waves propagate vertically. Because of the predominance of P-wave in the vertical response and S-wave in the horizontal response, one cycle of function $\phi(z)$ is relatively close to those of the body waves propagating vertically. Figs. 6 and 7 show the variations of the parameters, N and K, with frequency for vertical and horizontal responses, respectively. The trends in the figure indicate that, as the frequency increases, the shear beam in the further simplified model plays a less significant role but the springs play a more significant role through the imaginary part of the stiffness. This imaginary part of the spring is due to the radiation damping associated with wave propagation to the lateral infinity.

    A massless rigid foundation on a visco-elastic stratum underlain by a rigid base is considered next. The thickness of the stratum is assumed to be 2b, and $D = 2\%$ and $\nu = 1/3$ are assumed for soil. Dynamic responses of the foundation are computed by using the above obtained simple expressions. Fig. 8 shows the computed compliance function (flexibility ~ a_0 relationship) of this foundation for the vertical response. The results computed by a more rigorous approach (Tassoulas 1981) are also shown by dots. Again, good agreements between the two computed results are seen in the figure. The value $\overline{\alpha}$ controls the function $\phi(z)$. Figs. 9 and 10 show respectively the variation of $\overline{\alpha}$ with frequency and $\phi(z)$ at various frequencies. The fundamental natural frequency of the stratum is $a_0 = 0.785$ for one-dimensional S-wave propagation. As the frequency increases beyond this frequency, the wave starts to propagate to the lateral infinity. Thus the imaginary part of $\overline{\alpha}$ suddenly increases with a_0 as seen in Fig. 9. The resonance in compliance function in Fig. 10 shows at the frequency higher than this frequency. This is because the resonance in the vertical vibration is controlled by the P-waves and thus it is located around the natural frequency of the stratum in one-dimensional P-wave propagation.

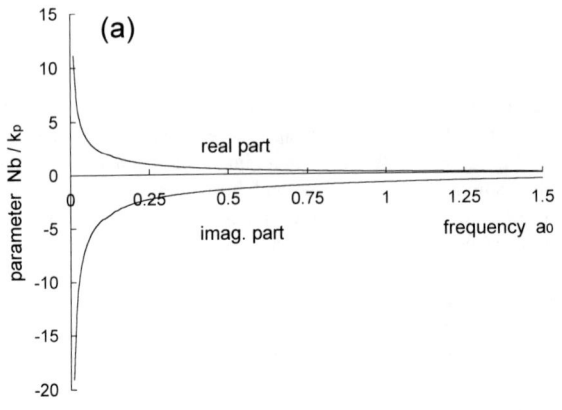

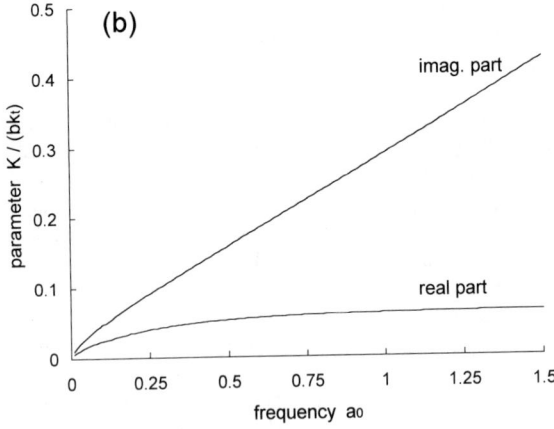

Fig. 6 Parameters N and K in Vertical Response of Foundation on Half-Space:
(a) Parameter N; (b) Parameter K

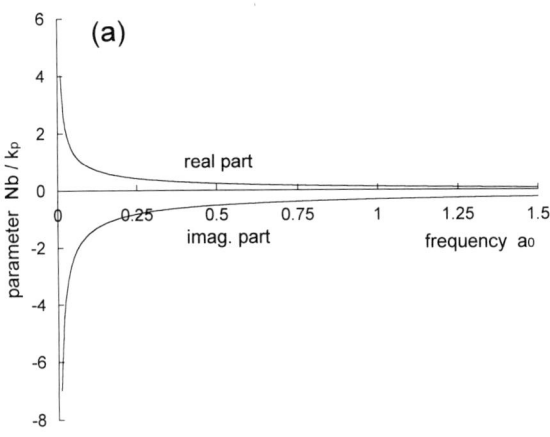

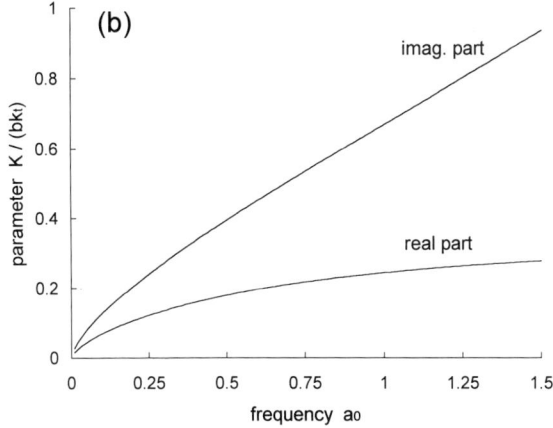

Fig. 7 Parameters N and K in Horizontal Response of Foundation on Half-Space:
(a) Parameter N; (b) Parameter K

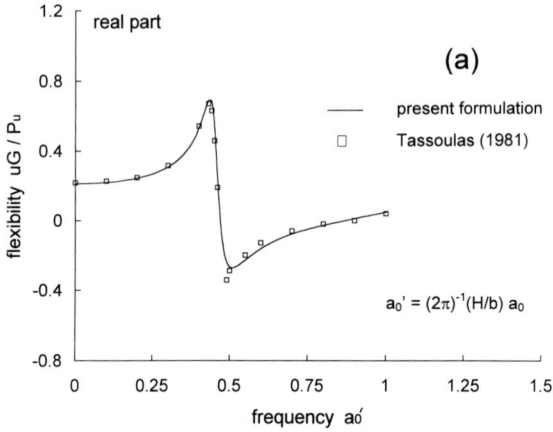

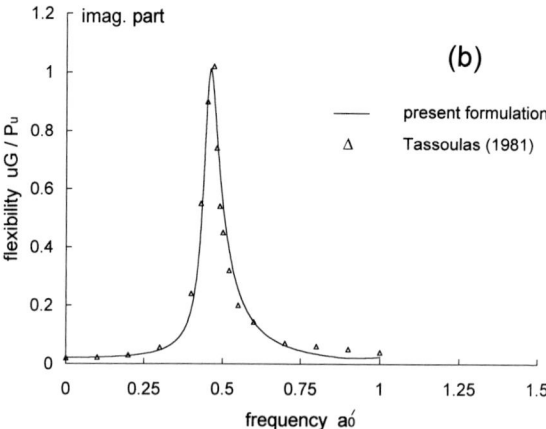

Fig. 8 Compliance Function for Vertical Repose of Foundation on Stratum:
(a) Real Part; (b) Imaginary Part

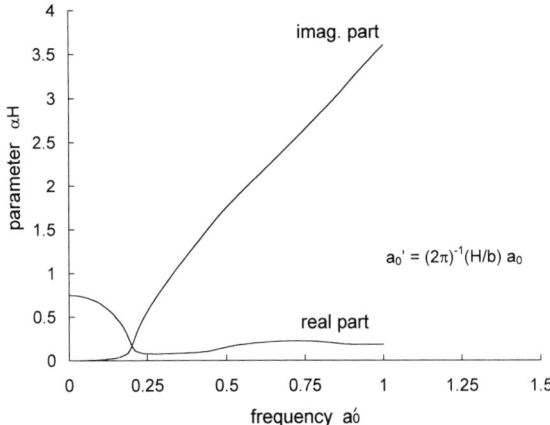

Fig. 9 Variation of α with Frequency for Vertical Response of Foundation on Stratum

Conclusions

Galerkin's method for weighted residual is applied to the previously proposed simplified ground model for the ground subjected to the load at the surface. This procedure leads the governing equation for the soil-foundation interaction behavior only along the ground surface, which enables us to model the ground surface behavior by a further simplified model. Solving this equation, the dynamic responses of a massless rigid foundation are formulated in simple explicit forms. Despite their significant simplicity, the developed formulations can predict the responses computed by far more complex rigorous formulations and can well capture the important dynamic characteristics of foundation responses which result from the wave propagation in a ground medium.

Appendix I. Development of Expressions

The soil domain is divided into three regions, $-\infty \le x \le -b$, $-b \le x \le b$ and $b \le x \le \infty$, in which the foundation occupies the second region and the first and second regions are outside of this region. The displacements in the regions outside the foundation are obtained by solving Eq. 11 with $p(x) = 0$ and imposing the displacement compatibility between the foundation and outside regions ($x = \pm b$). Thus, when the foundation displacement is u in translation and θ in rocking, the displacements along the ground surface are

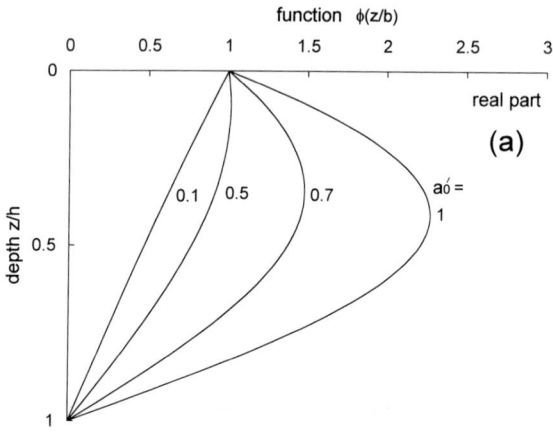

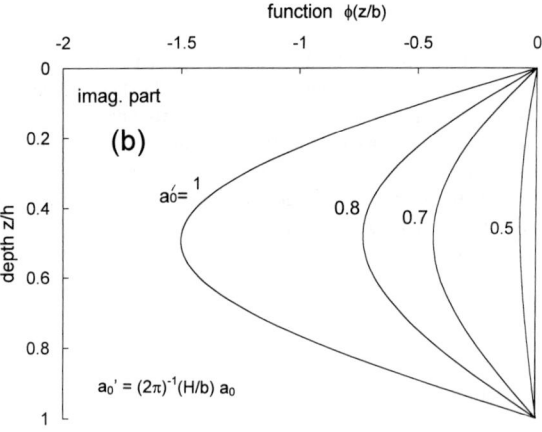

Fig. 10 Function φ(z) for Vertical Response of Foundation on Stratum at Various Frequencies: (a) Real Part; (b) Imaginary Part

translational responses

$$u(x) = \begin{cases} ue^{\beta(x+b)} & -\infty \le x \le -b \\ u & -b \le x \le b \\ ue^{-\beta(x-b)} & b \le x \le \infty \end{cases}$$

(29a)

rocking response

$$u(x) = \begin{cases} -b\theta e^{\beta(x+b)} & -\infty \le x \le -b \\ x\theta & -b \le x \le b \\ b\theta e^{-\beta(x-b)} & b \le x \le \infty \end{cases}$$

(29b)

where

$$\beta^2 = \frac{K - \omega^2 M}{N}$$

(30)

The loads applied to the foundation are carried by the reaction at the base and the reactions at the $x = \pm b$ from the outside of the base. The reaction at the base is $p(x)$ and the reactions from the outside are $Nu'(x)|_{x=-b}$ and $-Nu'(x)|_{x=b}$. Therefore, the equilibrium conditions of the foundation are stated as

$$P_u = \int_{-b}^{b} p(x)dx + Nu'(x)|_{x=-b} - Nu'(x)|_{x=b}$$

(31a)

$$P_\theta = \int_{-b}^{b} xp(x) + xNu'(x)|_{x=-b} - xNu'(x)|_{x=b}$$

(31b)

where P_u and P_θ = amplitudes of applied translational force and moment, respectively. Substituting $p(x)$ in Eq. 11 and $u(x)$ in Eq. 29 into Eq. 31, the following simple expressions are obtained for the foundation responses

$$P_u = 2\{b(K - \omega^2 M) + N\beta\}u \qquad \text{translational responses}$$

(32a)

$$P_\theta = 2\left\{N\beta b^2 + \frac{b^3}{3}(K - \omega^2 M)\right\}\theta \qquad \text{rocking response}$$

(32b)

Appendix II. References

Awojobi, A.O. (1972). "Vertical Vibration of a Rigid Circular Body and Harmonic Rocking of a Rectangular Body on an Elastic Stratum." Int. J. Solids Structures, 8, 579-774.

Dominguez, J. and Roesset, J.M. (1978). "Dynamic Stiffness of Rectangular Foundation." Report No. R78-20, MIT, Cambridge, Massachusetts.

Karasudhi, P., Keer, L.M., and Lee, S.L. (1968). "Vibratory Motion of a Body on an Elastic Half Plane." J. Applied Mechanics, 35, Trans. ASME, 90, Series E, Dec., 1-9.

Kausel, E., and Roesset, J.M. (1975). "Dynamic Stiffness of Circular Foundations." J. Engineering. Mechanics Div., ASCE, 103(4), 569-588.

Luco, J.E. (1976). "Vibrations of a Rigid Disc on a Layered Viscoelastic Medium." Nuc. Eng. Des., North-Holland Publishing Company, 36, 325-340.

Lysmer, J., Udaka, T., Seed, B.H., and Hwang, R. (1976). "FLUSH - A Computer Program for Approximate 3D Analysis of Soil Structure Interaction Problems." Rep. EERC 75/30, University of California, Berkeley.

Nogami, T. and Leung, M.B. (1990). "Simplified Mechanical Subgrade Model for Dynamic Response Analysis of Shallow Foundations." Int. J. Earthq. Engrg. Struct. Dyn., 19, 1041-1055.

Nogami, T. Zhu, J.X., and Itoh, T. (1992). "First and Second Order Dynamic Subgrade Models for Dynamic Soil-Pile Interaction Analysis." Geotechnical Special Technica Publication on Piles under Dynamic Loads, ASCE, No. 34, 187-206.

Nogami, T. (1996). "Simplified Subgrade Model for Three-Dimensional Soil-Foundation Interaction Analysis." Int. J. Soil Dyn. Earthq. Engrg., 15, 419-429.

Novak, M. (1974). "Dynamic Stiffness and Damping of Piles." Canadian Geotech. J., National Research Council of Canada, 11(4), 574-598.

Pak, R.Y.S. and Gobert (1991). "Forced Vertical Vibration of Rigid Discs with Arbitrary Embedment," J. Engineering. Mechanics Div., ASCE, 117(11), 2527-2548.

Tassoulas, J.L. (1981). "Elements for the Numerical Analysis of Wave Motion in Layered Media." Report No. R81-2, MIT, Cambridge, Massachusetts.

Aspects of Dynamic Centrifuge Testing of Soil-Pile-Superstructure Interaction

Daniel W. Wilson[1], Ross W. Boulanger[1], Bruce L. Kutter[1], and Abbas Abghari[2]
Members ASCE

Abstract

Results of dynamic centrifuge tests of pile-supported structures in soft or liquefied soils on the large centrifuge at U.C. Davis are used to evaluate several aspects of the physical modeling system. Aspects that are evaluated include model uniformity, input motion characteristics, test repeatability, pore fluid effects, vertical motions, and model container effects.

Introduction

Dynamic centrifuge tests of pile-supported structures in soft or liquefied soils were performed on the large centrifuge at U.C. Davis. These centrifuge experiments were among the first performed using the recently completed shaking table, and thus it was necessary to evaluate the centrifuge modeling system before analyzing the model structures. The importance of characterizing the centrifuge modeling system was demonstrated by the VELACS cooperative study (e.g., Arulanandan et al. 1994) and further discussed by Scott (1994). Difficulties or limitations with dynamic centrifuge modeling systems can include: (1) non-repeatability of model tests; (2) undesirable vertical motions; (3) input motions lacking the broad frequency content of real earthquake motions; and (4) container effects. These and other aspects of the dynamic centrifuge modeling system on the large centrifuge at U.C. Davis are evaluated using the results of the soil-pile-superstructure interaction experiments. Quantifying these modeling limitations was considered essential before using the data to evaluate seismic design methodologies for pile-supported structures.

---

[1] Graduate Student, Assistant Professor, and Professor, respectively, Department of Civil and Environmental Engineering, University of California, Davis, CA 95616
[2] Senior Materials and Research Engineer, Office of Structural Foundations, California Department of Transportation, Sacramento, CA 95819

Description of the Centrifuge and Model Layouts

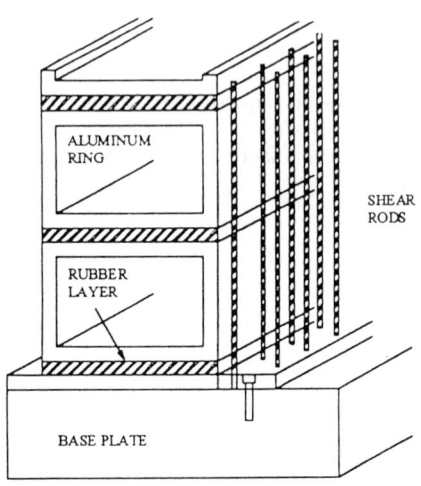

Figure 1: Schematic of rings and shear rods (from Divis et al. 1997)

The National Geotechnical Centrifuge at UC Davis has a radius of 9 m and is equipped with a large shaking table driven by two servo-hydraulic actuators (Kutter et al. 1994). Models were tested in a Flexible Shear Beam (FSB1) container having inside dimensions of 1.72 m long, 0.70 m deep, by 0.685 m wide. FSB1 consists of six hollow aluminum rings separated by 20 durometer neoprene. The mass of each of the upper three rings is about one-half the mass of each of the lower three rings. The combined mass of the six rings is about 25% of the soil profile mass (assuming the container is full of soil). The amount of neoprene separating the rings is varied such that the shear stiffness of the container increases with depth. The shear stiffness of the neoprene also varies with strain level (Stewart et al. 1997), such that the fixed base natural frequency of the empty container is about 15-20 Hz for the larger shaking events presented herein. Complementary shear stresses are provided by shear rods at each end of the container (Divis et al. 1997). A schematic of the rings, neoprene layers, and shear rods is shown in Figure 1.

Five containers with several pile-supported structures were each subjected to several earthquake motions. Full details for all five centrifuge tests can be found in Wilson et at. [1997 (a-e)]. All tests were performed at a centrifugal acceleration of 30 g, and results are presented in prototype units unless otherwise noted. For details of the applicable scaling laws, see Kutter (1995). A typical arrangement of structures and instrumentation is shown in Figure 2. In all cases, the soil profile consisted of two horizontal soil layers, as summarized in Table 1. The lower layer for all tests was dense Nevada sand, a fine, uniform sand (C_u=1.5, D_{50}=0.15mm). The upper layer was medium-dense Nevada sand in Csp1 and 3, loose Nevada sand in Csp2, and normally consolidated (NC) reconstituted Bay Mud (LL≈90, PI≈40) in Csp4 and 5. In all tests the sand was air pluviated, flushed with carbon dioxide, and saturated under vacuum. Pore fluid was water or a hydroxy-propyl methyl-cellulose (HPMC)-water mixture having a viscosity ten times that of water alone. Saturation was confirmed by measuring p-wave velocities before testing.

Table 1 - Summary of Soil Profiles

Container	Soil profile		Pore
	Upper layer (9.1 m thick)[a]	Lower layer (11.4 m thick)	fluid
Csp1	Sand ($D_r \approx 55\%$)	Sand ($D_r \approx 80\%$)	Water
Csp2	Sand ($D_r \approx 35\text{-}40\%$)	Sand ($D_r \approx 80\%$)	HPMC-water
Csp3	Sand ($D_r \approx 55\%$)	Sand ($D_r \approx 80\%$)	HPMC-water
Csp4 & 5	Reconstituted Bay Mud (NC)	Sand ($D_r \approx 80\%$)	Water

[a]Upper layer was only 6.1 m thick (prototype) in Csp4 & 5.

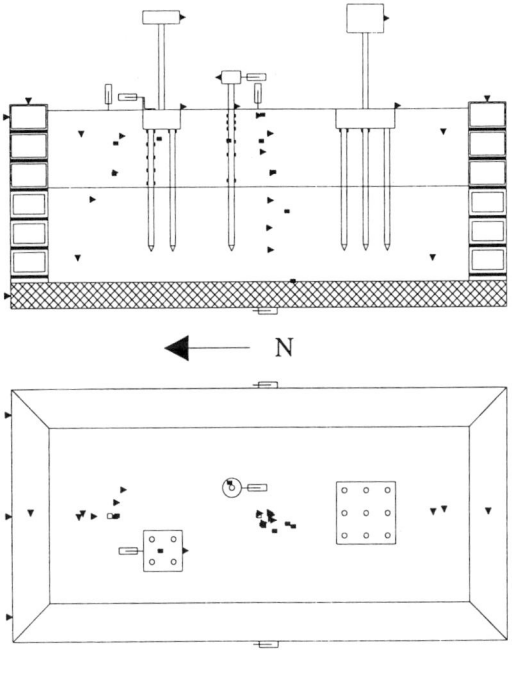

- Pore pressure   ·· Bending/Axial gage
⊏ displacement   ► Accelerometer

Figure 2: Typical model layout Csp1-5 (not all instruments shown)

Uniformity of Sand Layers

The density, uniformity, and repeatability of sand layers were evaluated by measuring the force required to push a 6 mm diameter rod with a 60° conical tip at various locations while at 1 g (Divis et al. 1994). The force was divided by the tip area for presentation as a penetration resistance (Q), although it is noted that Q reflects both tip and shaft resistances. Results of penetration tests on Csp3, after the pile groups were driven into the sand at 1 g, are shown in Figure 3. Tests in the free field showed nearly uniform profiles of Q, with Q being much higher in the lower dense layer than the upper loose layer. Three tests were located alongside the 2x2 and 3x3 pile groups, and these showed substantial increases in Q due to pile driving. Two tests were pushed between the piles of the 2x2 and 3x3 groups (through holes in the caps), and these showed even greater values of Q, particularly in the 3x3 group. Interpretation of these penetration tests is complicated by the low confining pressures (at 1 g), the mix of shaft and tip resistances, the relatively large zone of influence of the tip (e.g., 10-20 probe diameters is 9-18% of the total soil thickness), and the influence of the boundaries. Nonetheless, these data are a valuable indicator of specimen density and uniformity, and were useful for evaluating the pile installation effects.

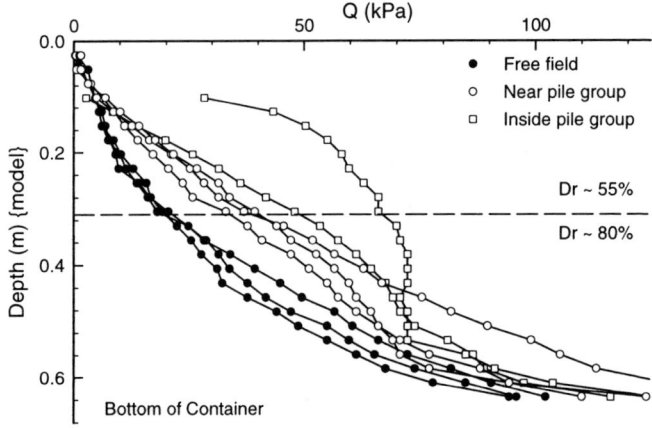

Figure 3: Results of penetration tests Csp3

Input Motions

Two important aspects of centrifuge input motions are: (1) the input motion should contain a reasonably full spectrum of earthquake frequencies for realistic dynamic modeling of pile-supported structures; and (2) the frequency content of the input motion should be reasonably unchanged when scaling the acceleration magnitude to minimize difficulties with evaluating nonlinear behavior between scaled shaking events. Note that gaps in the frequency content of input motions are a limitation of many dynamic centrifuge tests (Scott 1994).

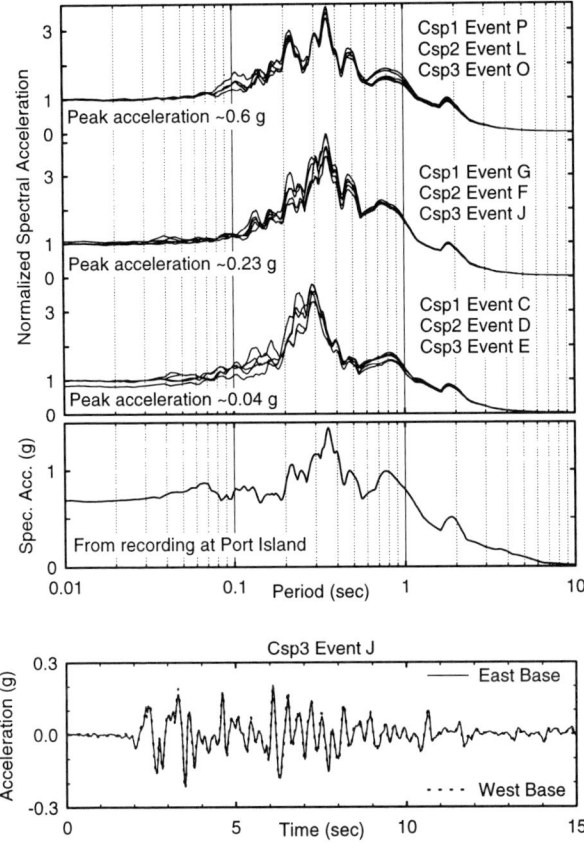

Figure 4: Repeatability of input motions
(all spectra at 5% damping)

The performance of the shaking table is shown in Figure 4. Each container
was shaken with several simulated earthquake events, each being a scaled version of a
record prepared by integrating and filtering strong motion records from Port Island
(83 m depth) in the Kobe Earthquake or Santa Cruz in the Loma Prieta Earthquake.
Acceleration response spectra (ARS, 5% damping) of the east and west base input
motions recorded during three scaled Kobe events ($a_{max} \approx 0.04$, 0.23, and 0.6 g
prototype) on each of three containers (total of nine events) are shown in Figure 4,
with the ARS normalized to a zero period value of one on the east actuator. The ARS
are very similar at each level of shaking, and show only small spectral variations

across the full operational range of the shaker (i.e., $a_{max} \approx 0.04$ to 0.6 g prototype corresponds to $a_{max} \approx 0.12$ to 18 g model for a centrifugal acceleration of 30 g). East and west base motions are also seen to be closely in-phase and parallel, as shown at the bottom of Figure 4 by the nearly identical acceleration time histories during a typical Kobe event. The ARS of the original recording from Port Island is shown in Figure 4 for comparison. The base motions retain the full frequency spectrum of the original recording in the range of interest (0.5-5 Hz prototype in this study), with the differences at higher and lower frequencies partially due to low and high pass filtering performed in creating the centrifuge input motion.

Comparison of Different Tests

The effect of changing pore fluid viscosity was evaluated by comparing results from Csp1 and Csp3 for four comparable shaking events. These containers had identical soil profiles and an identical single-pile-supported structure [model details in Wilson et al. 1995, Boulanger et al. 1997, Wilson et al. 1997 (a) and (c)], but the viscosity of the pore fluid differed by a factor of 10 (Table 1). The responses of the soil profile and single-pile-supported structure were very similar for comparable shaking events except as follows. The rate of pore pressure dissipation was alsways faster in Csp1 than in Csp3, as illustrated in Figure 5 by pore pressure time histories at similar locations during a Kobe event with $a_{max,base} \approx 0.23$ g. During shaking, however, the pore pressures increased at similar rates and underwent similar rapid changes. ARS for various locations in the upper sand layer in Csp1 and Csp3 also had similar normalized shapes, although Csp1 consistently had slightly greater spectral accelerations near a period of one second and slightly lower spectral accelerations at shorter periods. It was noted that comparisons of individual measurements for one comparable shaking event often showed slight differences, but comparisons of several instruments for several events were required to evaluate whether such slight differences followed a consistent trend. The slight differences between the responses in Csp1 and Csp3 were likely due to a combination of factors, including slight variations in soil densities, model preparation techniques, input motions, and the different pore fluids used.

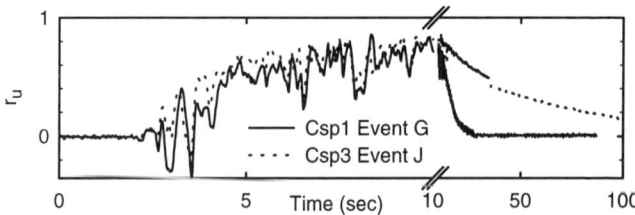

Figure 5: Effect of pore fluid viscosity on generation and dissipation of pore pressure in Csp1 (water) and Csp3 (HPMC-water)

The response of the single pile system during these same events is illustrated by bending moment time histories at depths of 3.8 and 5.3 m in Figure 6. Bending moments were normalized by the peak superstructure acceleration because the peak superstructure acceleration in Csp1 was about 50% greater than in Csp3. The difference in superstructure accelerations is due to both a 20% difference in the peak base input motion and the previously described differences in the soil profile ARS at the natural period of the structure (about one second). Normalized bending moments for this single pile system in Csp1 and Csp3 show very little difference during shaking, but do show interesting, although inconsequential, differences developing after shaking because of the different dissipation rates for excess pore pressures. Comparing bending moment time histories at other depths and other levels of shaking also gave very similar results, and thus the bending moment distributions at any time were essentially the same in Csp1 and Csp3. These results suggest that changing pore fluid viscosity by a factor of 10 had only minor effects on the soil-pile interaction.

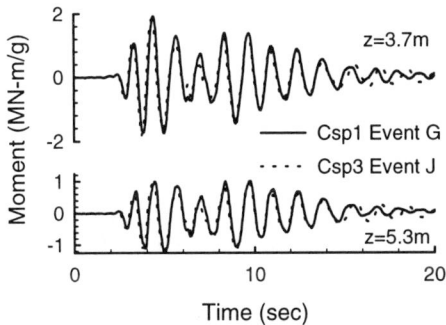

Figure 6: Effect of pore fluid viscosity on
bending moments (pile diameter D=0.67m)

For contrast, the effect of the upper soil layer on structural response and bending moment distribution is illustrated by a comparison of results from Csp2, 3, and 4 (Table 1). In Figure 7(a), the bending moment distributions versus depth are shown for an identical single pile system in Csp2, 3, and 4 during a Kobe event with $a_{max,base} \approx 0.23$ g. In Figure 7(b), these bending moments are normalized to a ground surface moment of unity. Note that liquefaction was more extensive in Csp2 than in Csp3 during these events, as evidenced by pore pressure time histories showing that pore pressures increased much quicker, and dissipated slower, in the $D_r \approx 35\%$ sand layer of Csp2 than in the $D_r \approx 55\%$ sand layer of Csp3 (Figure 8). The looser condition of the upper layer in Csp2, relative to Csp3, resulted in generally lower ground surface accelerations, lower peak superstructure accelerations, and a greater apparent softening of the liquefied soil's p-y resistance (Boulanger et al. 1997); these aspects of behavior are shown by the smaller bending moment at the ground surface [Figure 7(a)], but a greater depth to peak bending moment [Figure 7(b)]. In Csp4 (upper layer of soft clay), the peak superstructure acceleration was lower than that of Csp2 and 3, and the depth to peak bending moment was comparable to that of Csp2. These data are consistent with the expected effects of soil conditions on site response and soil-pile-superstructure interaction for the input motion used in these tests.

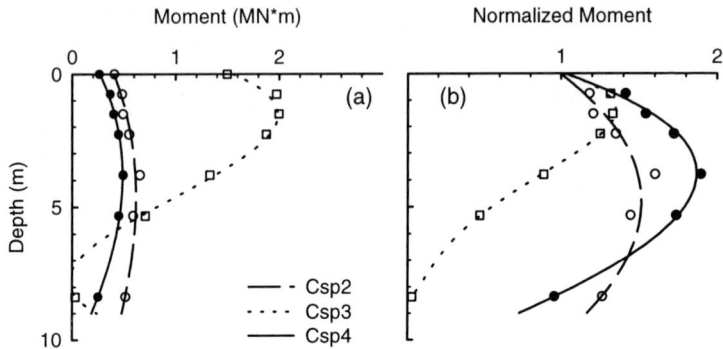

Figure 7: Bending moment distribution with varying soil types for Kobe events with $a_{max,base} \approx 0.23g$

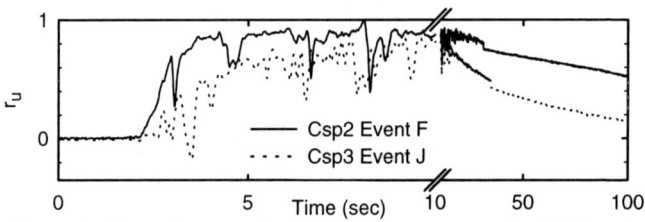

Figure 8: Generation and dissipation of pore pressure in 35% D_r soil (Csp2) and 55% D_r soil (Csp3)

Behavior of the Container and Soil Column System

The dynamic characteristics of a model container and its interaction with the soil column must be clearly understood if reliable interpretations of test results are to be made. Container effects on the soil column response have been studied using several different measurements of response (e.g., Fiegel et al. 1994, Van Laak et al. 1994, Whitman and Lambe 1986). In this study, the interaction is evaluated in terms of the coherency of horizontal motions and differential vertical displacements in the soil near the container ends.

The coherency of horizontal motions across the soil column and container rings indicates whether the container and soil are moving in unison during shaking. To measure coherency, accelerometers were attached to the individual rings of the FSB1 container and at corresponding depths near the center and corners of the soil profile. Accelerometer records were high-pass filtered and double integrated to get displacements. The accuracy and reliability of these procedures were demonstrated by placing accelerometers on opposite sites of displacement transducers and comparing the calculated and recorded relative displacement time histories. Results for several shaking events on each container show that horizontal acceleration and displacement time histories are nearly identical (i.e., highly coherent) at any given elevation in the soil column and on the corresponding container ring for tests involving nonliquefied sand or low shaking levels with soft clay.

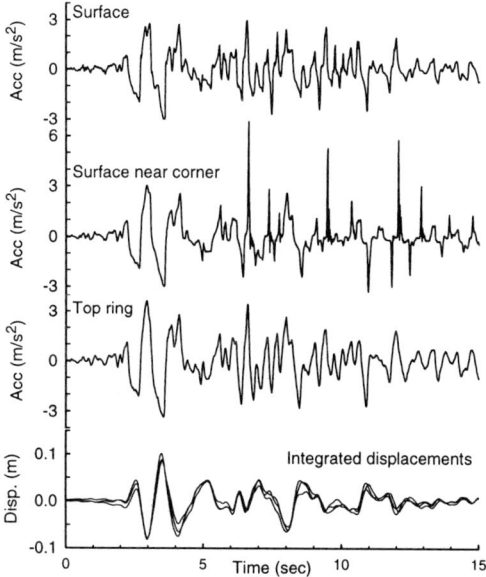

Figure 9: Coherency of motion near top of container
Csp1 Event G (Kobe, $a_{max,base} \approx 0.23g$)

Horizontal motions at shallow depths in Csp1 during a Kobe event ($a_{max,base} \approx 0.23$ g) causing liquefaction of the $D_r \approx 55\%$ layer late in shaking are shown in Figure 9. Accelerations at the soil surface near the center and one end of the model, and on the top ring, are seen to have similar waveforms but with differing high frequency contents later in shaking. In particular, several large high-frequency acceleration spikes were recorded near the end of the container. However, horizontal displacements relative to the container base at these three points were relatively uniform (bottom of Figure 9).

Spikes in acceleration records from centrifuge tests with liquefied soils have been observed by several investigators, while they have been less obvious in field data. In our tests, acceleration spikes have been observed throughout liquefied layers, near the middle and ends of the container, and in horizontal and vertical directions. Acceleration spikes have not been observed when the excess pore pressure ratio is less than about 70%. Acceleration spikes coincide with rapid pore pressure drops, and thus are likely due to the uniform soil profile "locking" up all at once as the sand goes through a phase transformation (i.e., the transition from contractant to dilatant behavior). Additional research, however, is needed to evaluate potential instrumentation effects, such as a local interaction between the instrument and liquefied soil. It should also be noted that these high frequency acceleration spikes have a lesser effect on the velocities, displacements and kinetic energy in the soil profile and structural models.

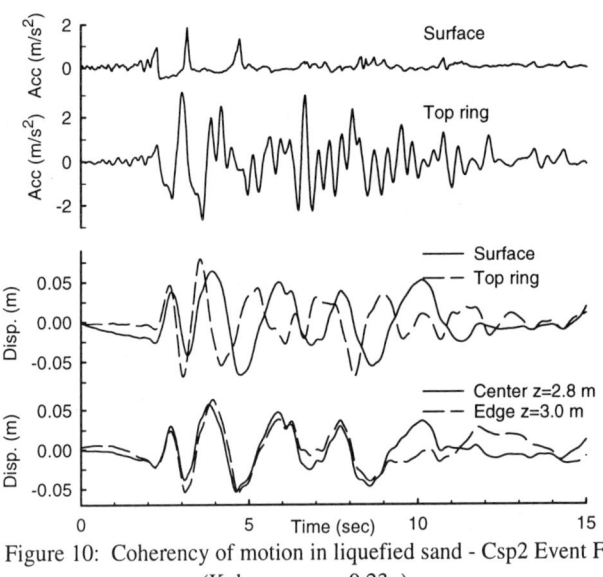

Figure 10: Coherency of motion in liquefied sand - Csp2 Event F
($a_{max,base} \approx 0.23$g) (Kobe,)

Horizontal motions at shallow depths in Csp2 during a Kobe event ($a_{max, base} \approx 0.23$ g) causing liquefaction of the $D_r \approx 35\%$ layer are shown in Figure 10. Accelerations at the surface of the soil near the center of the container were very different than the acceleration of the top ring. Furthermore, the displacements of the top ring and the soil relative to the container base were very different, at times nearly 180° out of phase.   In this case, when r_u was high and D_r was low, the soil column became much softer than the container, as shown by the predominant frequency content of recorded motions in the profile and on the container. As a result, the container restricted lateral movements near its edge. While this is not ideal, it is physically difficult to avoid and may be incorporated into numerical analyses if necessary. Coherency of horizontal motions, however, improved with depth in the liquefied soil layer. This is illustrated at the bottom of Figure 10 by the horizontal displacements relative to the container base for two locations at the same elevation deeper in the liquefied layer.

A similar set of plots from Csp4 for a Kobe event ($a_{max,base} \approx 0.23$ g), where the upper soil was normally consolidated clay, are shown in Figure 11. In this case, however, the top ring was empty and the soil surface was level with the second ring. The upper plot shows the difference between displacements at the surface center of the soil profile and the second ring relative to the container base. The lower plot, however, shows that by the third ring, the container and the soil are moving mostly together.

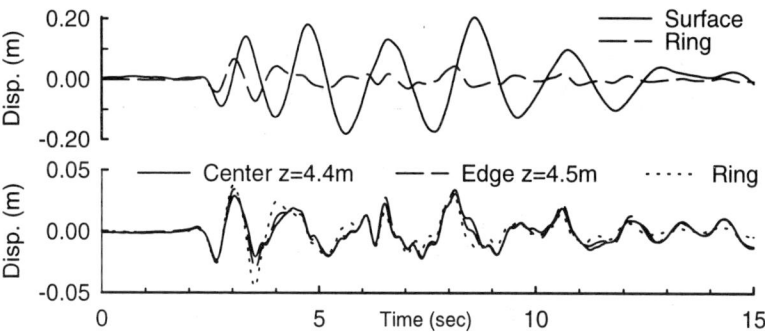

Figure 11: Coherency of motion in clay - Csp4 Event D
(Kobe, $a_{max,base} \approx 0.23$g)

In addition to the container moving with the soil, the soil profile should also deform in shear as opposed to column bending. In shear, there is no vertical strain when the soil profile deforms horizontally, while column bending will cause one side to compress and the other to extend. The container should help minimize column bending by providing complementary shear stresses at the end interfaces between the soil and the container. Discussions of the role of complementary shear stresses and rocking in centrifuge modeling can be found in Whitman and Lambe (1986) and Scott (1994).

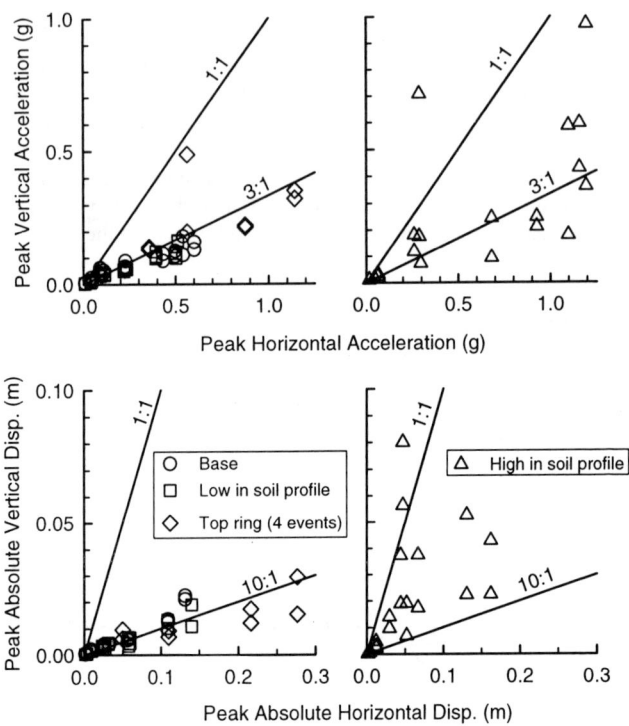

Figure 12: Peak vertical vs. peak horizontal accelerations and
displacements throughout model

In Csp2 ($D_r \approx 35\%$ upper layer), vertical accelerometers were included at the
north and south ends of the model container base and top ring, and near the bottom
and top of the soil profile (total eight transducers), in order to quantify rocking of the
container and soil column. Figure 12 is a summary plot of the recorded peak
accelerations and integrated peak absolute displacements from these transducers. The
peak vertical accelerations were typically 20 to 30% of the peak horizontal
accelerations at all locations other than the upper soil profile, and the peak vertical
displacements were typically less than about 10% of the peak horizontals, again other
than in the upper soil profile. Note that these data are for the ends of the container,
while vertical motions within the central portion of the container are expected to be
much smaller. These data show that the shaking table and FSB1 container do not
introduce significant rocking or pitching motions, and that the lower halves of the soil
profile have similarly low levels of vertical motion. However, the vertical

accelerations and displacements in the upper soil profile are comparable to their horizontal counterparts, which would seem to indicate very poor performance within the soil. Note, however, that liquefaction occurred in all but the smallest events in Csp2. For the smallest events, the vertical accelerations and displacements recorded in the upper soil profile appear to be consistent with the magnitude of the other vertical recordings. This indicates that the large vertical motions near the ends of the upper soil profile were due to liquefaction effects.

There are at least two simple modes of vertical displacement for the soil near the ends of the container. In the first scenario, the soil profile deforms as a column, as shown schematically in Figure 13(a). As shown, when the base accelerates to the north, the relative displacement of the surface will be to the south. The vertical displacement at the south end of the container will be down, and the opposite will occur at the north end. In the second scenario, the liquefied soil "sloshes" in a relatively rigid container, as shown schematically in Figure 13(b). Here, when the base accelerates to the north and the relative displacement of the surface is to the south, the soil at the south end of the container is likely to slosh upward, and the vertical displacement at the north end will be negative.

    (a) Column bending   (b) Sloshing

Figure 13: Two possible modes of deformation (greatly exaggerated)

Vertical and horizontal displacements of the ground surface relative to the container base in Csp2 ($D_r \approx 35\%$ upper layer) are summarized in Figure 14 for both (a) a non-liquefaction and (b) a liquefaction event. The convention used in Figure 14 is that positive horizontal displacement is a displacement to the south, and positive vertical displacement is upward. In Figure 14(a), the horizontal displacement and the north vertical displacement are nearly in phase, while the south vertical is nearly 180° out of phase. This is consistent with the expected behavior for bending of the soil column, as previously described. Also, the vertical displacements are on the order of 10% of the horizontals. In Figure 14(b), the opposite phasing occurs, with the south vertical moving in phase with the horizontal and the north nearly 180° out. This would be consistent with the sloshing mode previously described, and is consistent with Figure 10 where we saw the effects of the soil profile becoming softer than the container. Also, in Figure 14(b) we see that the vertical displacements are of the same magnitude as the horizontals. Although this is a greatly simplified analysis, it is consistent with the argument that the large vertical displacements are not due to a lack of complementary shear stresses, but due to the difference in stiffness between the

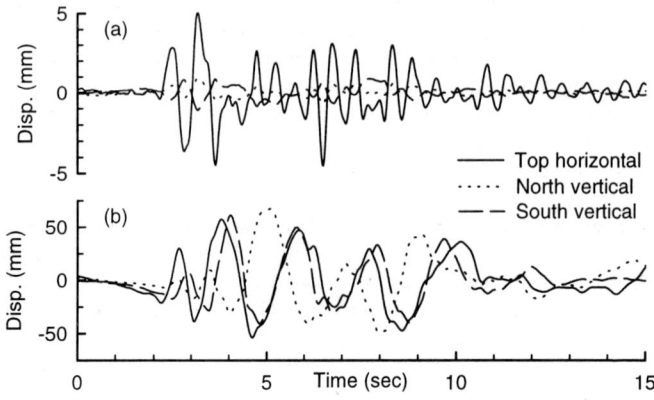

Figure 14: Relative displacements at the top of the FSB1 container
showing (a) column bending, and (b) sloshing of liquefied soil

soil and the container. While not ideal, an appreciation of this limitation is necessary
for realistic interpretations or analyses of the centrifuge data.

Pore Pressures Near Structures

The influence of the pile-supported structures on the excess pore pressures in
the upper sand layer of Csp1-3 was evaluated by placing pore pressure transducers
both near and far from the structures. Figure 15 shows excess pore pressure ratios (r_u)
at depths of 3.7 m in the $D_r \approx 35\%$ layer of Csp2: (a) at a "free field" location, (b)
about 0.3 m prototype from a single pile supported structure, and (c) about 3 m
prototype from a 2x2 pile group. Pore pressures near the single pile system show a
cyclic component at the predominant period of the single pile system (e.g., compare
Figures 8 and 15). Pore pressures near the 2x2 pile group also show a significant
cyclic component corresponding to the horizontal motions of the pile cap. While pore
pressures near structures are clearly not equal to those in the free field, these data
show very similar trends in their mean values over time.

Conclusions

Results from dynamic centrifuge tests of pile-supported structures in soft or
liquefied soils were used to evaluate several aspects of the centrifuge modeling
system that could potentially affect subsequent interpretations and analyses. Detailed

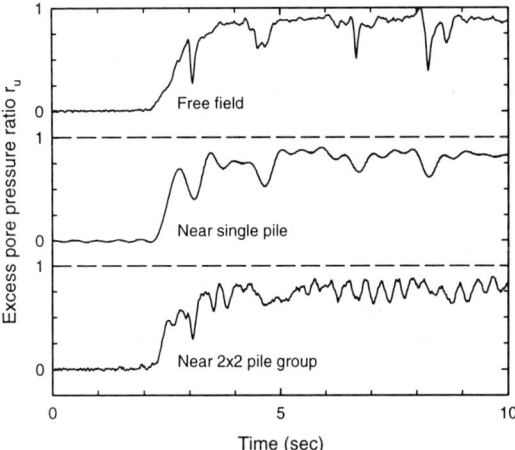

Figure 15: Excess pore pressure ratio in free field and
near structure

examination of the centrifuge modeling system was necessary because of the newness
of the shaking table, and since recent reviews have highlighted important limitations
that can exist in dynamic centrifuge systems (Scott 1994, Arulanandan et al. 1994).

Performance of the shaking table on the large centrifuge at UC Davis was
shown to be satisfactory. Full frequency spectra of desired input motions (including
real earthquake records) were recreated, with the motions being scaleable and
repeatable. Dynamic vertical displacements at the ends of the container base were
limited to about 10% of the dynamic horizontal displacements, indicating that rocking
of the container base was reasonably small over the full operating range of the shaker.

The FSB1 container produced satisfactorily uniform and coherent horizontal
motions, with relatively little rocking of the soil column, in tests on nonliquefied sand
or even liquefied D_r~55% Nevada sand. Incoherent horizontal motions and
differential vertical displacements developed at shallow depths in upper layers of
liquefied D_r~35% Nevada sand or strongly-shaken soft clay, indicating that the soil
column had become effectively "softer" than the FSB1 container in these tests.

Changing pore fluid viscosity by a factor of ten between two containers had
negligible effect on the soil-pile interaction, with or without liquefaction of the upper
soil layer. Furthermore, the nearly identical dynamic pore pressures and bending
moment distributions obtained in these two tests showed that reasonably repeatable
test results could be obtained nearly a year apart.

Bending moment distribution versus depth was found to be highly dependent on the soil stiffness. In tests with liquefied sand at low D_r and in tests with soft clay the maximum bending moment occurred much deeper than in tests with liquefied sand at higher D_r. The apparent p-y resistance of liquefied sand was shown to be strongly dependent on its D_r (Boulanger et al. 1997).

Acknowledgments

Support for this research was provided by Caltrans under contract number 65V495. The contents do not necessarily represent the policy of that agency nor endorsement by the State government. Design and construction of the shaker was supported by NSF, the Obayashi Corporation, and Caltrans. The authors would like to acknowledge the suggestions and assistance of I. M Idriss, D. P. Stewart, T. Kohnke, P. Robins, and C. Curras.

References

Arulanandan, K., Dobry, R., Elgamal, A.-W., Ko, H. Y., Kutter, B. L., Prevost, J., Riemer, M. F., Schofield, A. N., Scott, R. F., R. B. Seed, Whitman, R. V., and Zeng, X. (1994). "Interlaboratory studies to evaluate the repeatability of dynamic centrifuge model tests." *Dynamic Geotechnical Testing II, ASTM STP 1213*, R. J. Ebelhar, V. P. Drnevich, and B. L. Kutter, Eds., American Society for Testing and Materials, Philadelphia, pp. 400-422.

Boulanger, R. W., Wilson, D. W., Kutter, B. L., and Abghari, A. (1997). "Soil-pile-superstructure interaction in liquefiable sand." Preprint, 1997 Annual Transportation Research Board Meeting (TRB), Washington, D.C.

Divis, C. J., Kutter, B. L., Idriss, I. M., Goto, Y., and Matsuda, T. (1996). "Uniformity of specimen and response of liquefiable sand model in a large centrifuge shaker." *Proc., Sixth Japan-U.S. Workshop on Earthquake Resistant Design of Lifeline Facilities and Countermeasures Against Soil Liquefaction*, Hamada and O'Rourke, Eds., NCEER-96-0012, SUNY, Buffalo, pp. 259-273.

Fiegel, G. L., Hudson, M., Idriss, I. M., Kutter, B. L., and Zeng, X. (1994). "Effect of model containers on dynamic soil response." *Proc. Centrifuge 94*, Leung, Lee & Tan, Eds., Balkema, Rotterdam, pp. 145-150.

Kutter, B. L., Idriss, I. M., Kohnke, T., Lakeland, J., Li, X. S., Sluis, W., Zeng, X., Tauscher, R., Goto, Y., and Kubodera, I. (1994). "Design of a large earthquake simulator at UC Davis." *Proc. Centrifuge 94*, Leung, Lee & Tan, Eds., Balkema, Rotterdam, pp. 169-175.

Kutter, B. L. (1995). "Dynamic centrifuge modeling of geotechnical structures." *Transportation Research Record 1336*, TRB, National Research Council, Washington, D.C., pp. 24-30.

Scott, R. F. (1994). "Review of progress in dynamic geotechnical centrifuge research." *Dynamic Geotechnical Testing II, ASTM STP 1213*, R. J. Ebelhar, V. P. Drnevich, and B. L. Kutter, Eds., American Society for Testing and Materials, Philadelphia, pp. 305-329.

Stewart, D. P., Kutter, B. L., Higuchi, S., Robins, P. N., Narayanan, K. R., and
Thompson, D. J. (1997). "Centrifuge modeling of the seismic response of LNG
production facility structures: Phase I." Report No. UCD/CGM-97/02, Center
for Geotechnical Modeling, Department of Civil and Environmental Engineering,
University of California, Davis, March.

Van Laak, P. A., Taboada, V. M., Dobry, R., and Elgamal, A.-W. (1994).
"Earthquake centrifuge modeling using a laminar box." *Dynamic Geotechnical
Testing II, ASTM STP 1213*, R. J. Ebelhar, V. P. Drnevich, and B. L. Kutter,
Eds., American Society for Testing and Materials, Philadelphia, pp. 370-384.

Whitman, R. V., and Lambe, P. C. (1986). "Effect of boundary conditions upon
centrifuge experiments using ground motion simulation." *Geotechnical Testing
Journal*, ASTM GTJODJ, Vol. 9, No. 2, pp. 61-71.

Wilson, D. W., Boulanger, R. W., Kutter, B. L., and Abghari, A. (1995). "Dynamic
centrifuge tests of pile-supported structures in liquefiable sand." *Proc., National
Seismic Conference on Bridges and Highways*, Sponsored by Federal Highways
Administration and Caltrans, San Diego, CA, December 10-13.

Wilson, D. W., Boulanger, R. W., and Kutter, B. L., [1997 (a)]. "Soil-pile-
superstructure interaction at soft or liquefiable soil sites - Centrifuge data report
for Csp1." Report No. UCD/CGMDR-97/02, Center for Geotechnical Modeling,
Department of Civil and Environmental Engineering, University of California,
Davis, February.

Wilson, D. W., Boulanger, R. W., and Kutter, B. L., [1997 (b)]. "Soil-pile-
superstructure interaction at soft or liquefiable soil sites - Centrifuge data report
for Csp2." Report No. UCD/CGMDR-97/03, Center for Geotechnical Modeling,
Department of Civil and Environmental Engineering, University of California,
Davis, February.

Wilson, D. W., Boulanger, R. W., and Kutter, B. L., [1997 (c)]. "Soil-pile-
superstructure interaction at soft or liquefiable soil sites - Centrifuge data report
for Csp3." Report No. UCD/CGMDR-97/04, Center for Geotechnical Modeling,
Department of Civil and Environmental Engineering, University of California,
Davis, February.

Wilson, D. W., Boulanger, R. W., and Kutter, B. L., [1997 (d)]. "Soil-pile-
superstructure interaction at soft or liquefiable soil sites - Centrifuge data report
for Csp4." Report No. UCD/CGMDR-97/05, Center for Geotechnical Modeling,
Department of Civil and Environmental Engineering, University of California,
Davis, February.

Wilson, D. W., Boulanger, R. W., and Kutter, B. L., [1997 (e)]. "Soil-pile-
superstructure interaction at soft or liquefiable soil sites - Centrifuge data report
for Csp5." Report No. UCD/CGMDR-97/06, Center for Geotechnical Modeling,
Department of Civil and Environmental Engineering, University of California,
Davis, February.

Axial Vibrations of Circular Pile Foundations
in Layered Soil Medium

C.V. Girija Vallabhan[1], F ASCE

Abstract

Using minimum potential energy theorem and variational calculus, Vallabhan has developed a new model to compute the settlements of axially loaded piles placed in a layered soil medium. Here, using Hamilton's principle, the model is further extended to compute the dynamic response of piles placed in a multi-layered elastic soil medium. Perfect compatibility of displacements between the soil and the pile is assumed. The solution consists of assuming the vertical displacement of the soil as a product of the pile displacement and a mode function. Two coupled differential equations are developed which are solved numerically using an iterative procedur[1]e. The method has the advantage over other existing methods that a compatible displacement of the soil between soil layers is incorporated in the model. Overall, the model is relatively very simple.

Introduction

Piles, piers, also known as drilled caissons, form a very efficient type of foundation to transfer heavy concentrated loads to deeper soil medium. These foundations are often subjected to dynamic vibratory forces such as machine induced forces and earthquakes etc. The problem here concerns with the dynamic interaction that occur between the pile and the soil medium. Often engineers need to predict and control the maximum dynamic displacement response of the piles under working load conditions, in order to have a good performance of the overall system. Computation of the dynamic settlement of these piles is relatively difficult, especially when the soil is stratified. Many researchers have undertaken the study of this complex interaction problem, experimentally as well as

---

[1] Professor of Civil Engineering, Texas Tech University, Lubbock, Texas.

theoretically. Unfortunately theoretical solution becomes too complicated even with many simplified assumptions and one has to depend on charts and figures to come up with some solutions. Using energy principles, Vallabhan (1995) and Vallabhan and Mustafa (1995) have developed a very simple mathematical model to compute the axial displacement of piles subjected to axial loads in linear layered soil medium. Earlier, using Hamilton's principle, Vallabhan (1996) extended this model to compute dynamic response of piles in a two layer soil medium subjected to harmonic axial loads. This paper deals with an extension of the previous dynamic model to consider response of a pile in a multi-layered soil medium.

Brief Review of Past Research

One of the early models for computing the dynamic response of piles is to assume that the pile-soil system act as a spring underneath the machine foundation. This method requires the assessment of the soil-pile spring constant which is too simplified and empirical which may behave too far from reality. The well known Reese model (Seed and Reese, 1957), was used to compute this constant which essentially gives the static performance of piles. If the piles are supported on rigid rock, they are considered as bearing piles and Richart (1970) has given some frequency characteristics of these piles based on pile properties only. Nogami and Novak (1976) appear to be the first to formulate a theoretical model for the analysis of vertical vibration of piles. Later Novak (1977) using the Baranov's concept of horizontal propagation of stress waves through layered strips of soil developed a model for dynamic response of floating piles. The Baranov solution does not use the compatibility condition between the infinitesimally thin layers. Kuhlemeyer (1978) compared Novak's results using finite element solutions. Novak and El-Sharnouby (1983) presented charts and figures to compute the vertical response of friction piles, i.e., neglecting the effect of the soil at the bottom of the pile. Nogami, et al. (1992) subgrade models for dynamic soil-pile interaction analysis. A special publication by ASCE edited by Prakash (1992) summarized the work done so far on this topic.

Theory of the New Model

A horizontally stratified layered soil medium carrying a cylindrical pile is shown in Fig. 1. It is assumed that the pile and the surrounding soil have linear material properties and have perfect compatibility of displacements at the interface during the motion. The problem is analyzed using cylindrical coordinates (r, θ, z). E_p, A_p, ρ_p, R and l are the modulus of elasticity, area of cross section, mass density, radius and length of the pile respectively. Each layer of soil is further divided into thin layers in order to increase accuracy of the solution. A thin i-th soil layer of thickness h_i with its own local coordinate system is shown in Fig. 2. From practical considerations, for the case of axial (vertical) motion, it can be assumed that the radial displacement $\bar{u}(r,z,t)$ in the soil is negligible compared to the vertical displacement $\bar{w}(r,z,t)$ and thus, it is assumed that the displacements at any point (r,z) in the soil medium surrounding the pier are:

$$\bar{u}(r,z,t) = 0, \quad \bar{v}(r,z,t) = 0, \quad \text{and} \quad \bar{w}(r,z,t) = w(z.t).\phi(r) \qquad (1)$$

$\bar{v}(r,z,t) = 0$, because of symmetry with respect to the z- axis. Here, $w(z,t)$ represents the axial displacement of the pile and the soil column under the pile, while $\phi(r)$ is a mode function in the surrounding soil medium, where $\phi(r) = 1$ at the pile-soil interface and $\phi(r) = 0$ at $r = \infty$. The internal strains in the surrounding soil medium become:

$$\varepsilon_r = \varepsilon_\theta = \gamma_{r\theta} = \gamma_{z\theta} = 0, \text{ while}$$

$$\varepsilon_z = \frac{dw}{dz}.\phi(r) \quad , \text{ and}$$

$$\gamma_{rz} = w.\frac{d\phi}{dr} \qquad (2)$$

The corresponding non-zero stress components are:

$$\sigma_r = \sigma_\theta = \frac{\mu}{1-\mu}\bar{E}_i\varepsilon_z$$

$$\sigma_z = \bar{E}_i \, \varepsilon_z$$

and $\quad \tau_{rz} = G_i\gamma_{rz} \qquad (3)$

where $\quad \bar{E}_i = \dfrac{E_i(1-\mu_i)}{(1+\mu_i)(1-2\mu_i)}$, $\quad E_i, \mu_i, G_i$ and ρ_i are the values of Young's modulus, Poisson's ratio, shear modulii and mass density of the i-th soil layer. The displacement $w(z,t)$ is interpolated between the two end nodes of the i-th layer as

$$w(z,t) = w_1(1-\frac{z}{h_i}) + w_2\frac{z}{h_i} \qquad (4)$$

where w_1 and w_2 are the local displacements of the nodes of the i-th layer of the pile which are functions of time only. The total kinetic energy of the pile, T_{pi} and the soil, T_{si} for this layer is

$$T_{pi} + T_{si} = \frac{1}{2}\int_0^{h_i} A_p\rho_p \dot{w}^2 dz + \pi \int_R^\infty \int_0^{h_i} \rho_i \dot{w}^2 \phi^2 r dr dz \qquad (5)$$

Substituting Eq.(4) in the above equation,

$$T_{pi} + T_{si} = \frac{h_i}{12}\left(A_p\rho_p + 2\pi\rho_i\int_R^\infty \phi^2 r\,dr\right)\begin{Bmatrix}\dot{w}_1\\\dot{w}_2\end{Bmatrix}'\begin{bmatrix}2 & 1\\1 & 2\end{bmatrix}\begin{Bmatrix}\dot{w}_1\\\dot{w}_2\end{Bmatrix}$$

$$= \frac{m_i}{2}\begin{Bmatrix}\dot{w}_1\\\dot{w}_2\end{Bmatrix}'\begin{bmatrix}2 & 1\\1 & 2\end{bmatrix}\begin{Bmatrix}\dot{w}_1\\\dot{w}_2\end{Bmatrix}$$

(6)

where $m_i = \dfrac{h_i}{6}\left(A_p\rho_p + 2\pi\rho_i\displaystyle\int_R^\infty \phi^2 r\,dr\right)$. Here $\dot{w} = \dfrac{\partial w}{\partial t}$. Now m_i contains the soil

mass in the i-th layer which will vibrate in phase with the pile in that layer. The strain energies in the pile, U_{pi} and the soil, U_{si} of the i-th layer are

$$U_{pi} + U_{si} = \frac{1}{2}\int_0^{h_i} E_p A_p \varepsilon_z^2 dz + \frac{1}{2}\int_{vol}\left(\sigma_z\varepsilon_z + \tau_{rz}\gamma_{rz}\right) r\,dr\,d\theta\,dz$$

$$= \frac{1}{2}E_p A_p\int_0^{h_i}\left(\frac{dw}{dz}\right)^2 dz + \pi\int_R^\infty\int_0^{h_i}\left\{\overline{E}_i\left(\frac{dw}{dz}\phi\right)^2 + G_i\left(w\frac{d\phi}{dr}\right)^2\right\} r\,dr\,dz$$

$$= \frac{1}{2h_i}\left(E_p A_p + 2t_i\right)\begin{Bmatrix}w_1\\w_2\end{Bmatrix}'\begin{bmatrix}1 & -1\\-1 & 1\end{bmatrix}\begin{Bmatrix}w_1\\w_2\end{Bmatrix} + \frac{h_i}{12}k_i\begin{Bmatrix}w_1\\w_2\end{Bmatrix}'\begin{bmatrix}2 & 1\\1 & 2\end{bmatrix}\begin{Bmatrix}w_1\\w_2\end{Bmatrix}$$

$$= \frac{1}{2}\begin{Bmatrix}w_1\\w_2\end{Bmatrix}'\begin{bmatrix}\alpha_i & \beta_i\\\beta_i & \alpha_i\end{bmatrix}\begin{Bmatrix}w_1\\w_2\end{Bmatrix}$$

(7)

where, $k_i = 2\pi G_i\displaystyle\int_0^{h_i} r\left(\frac{d\phi}{dr}\right)^2 dr$, $2t_i = 2\pi\overline{E}_i\displaystyle\int_0^{h_i}\phi^2 r\,dr$ and

$\alpha_i = \dfrac{(E_p A_p)_i}{h_i} + \dfrac{k_i h_i}{3}$ and $\beta_i = -\dfrac{(E_p A_p)_i}{h_i} + \dfrac{k_i h_i}{6}$

Using Hamilton's principle for the i-th layer, and taking variations in w and ϕ and using principles of variational calculus, the following three field equations are obtained for the i-th layer.

$$m_i\begin{bmatrix}2 & 1\\1 & 2\end{bmatrix}\begin{Bmatrix}\ddot{w}_1\\\ddot{w}_2\end{Bmatrix} + \begin{bmatrix}\alpha_i & \beta_i\\\beta_i & \alpha_i\end{bmatrix}\begin{Bmatrix}w_1\\w_2\end{Bmatrix} = \begin{Bmatrix}f_1\\f_2\end{Bmatrix}$$

(8)

Here f_1 and f_2 are the internal forces at the nodes. Combining all the element matrices to obtain global behavior, we get the global matrix as

$$[M]\{\ddot{W}\} + [K]\{W\} = \{F\}$$

(9)

where the global mass matrix is given as

$$
M = \begin{bmatrix}
2m_1 & m_1 & & & & & \\
m_1 & 2(m_1 + m_2) & m_2 & & & & \\
& m_2 & 2(m_2 + m_3) & m_3 & & & \\
& & m_3 & 2(m_3 + m_4) & m_4 & & \\
& & & \cdot & \cdot & \cdot & \\
& & & & \cdot & \cdot & \cdot \\
& & & & & m_{n-1} & 2(m_{n-1} + m_n) & m_n \\
& & & & & & m_n & 2m_n
\end{bmatrix}
$$

(10)

and the global stiffness matrix is

$$
[K] = \begin{bmatrix}
\alpha_1 & \beta_1 & & & & & \\
\beta_1 & \alpha_1 + \alpha_2 & \beta_2 & & & & \\
& \beta_2 & \alpha_2 + \alpha_3 & \beta_3 & & & \\
& & \beta_3 & \alpha_3 + \alpha_4 & \beta_4 & & \\
& & & \cdot & \cdot & \cdot & \\
& & & & \cdot & \cdot & \cdot \\
& & & & \beta_{n-1} & \alpha_{n-1} + \alpha_n & \beta_n \\
& & & & & \beta_n & \alpha_n
\end{bmatrix}
$$

(11)

where n is the total number of degrees of freedom including the required number of soil layers underneath the pile. For the soil column under the pile for $l < z < \infty$, the components of the mass matrix can also be computed by the circular area enclosed by the pile multiplied by the respective density of the soil. and

$$
\{W\}^t = \{W_1 \quad W_2 \quad W_3 \quad W_4 \quad . \quad . \quad W_{n-1} \quad W_n\}^t.
$$

(12)

By taking variations in ϕ, we get the differential equation for the soil as,

$$
-\frac{d}{dr}\left(r \frac{d\phi}{dr}\right) + \gamma^2 r\phi = 0 \quad \text{for } R < r < \infty
$$

(13)

with $\phi = 1$ @ $r = R$ and $\phi = 0$ @ $r = \infty$.

$$\left(\frac{\gamma}{R}\right)^2 = \frac{-\sum_i \left(\rho_i \int_0^l \left(\frac{dw}{dt}\right)^2 dz + \overline{E}_i \int_0^\infty \left(\frac{dw}{dz}\right)^2 dz\right)}{\sum_i G_i \int_0^\infty w^2 dz} \tag{14}$$

The summation is for all the layers including the soil layers underneath the pile. The boundary conditions are: @ $z = 0$, we have a concentrated mass at the top $\widetilde{M}_0$ representing the footing and an applied vertical force $\widetilde{P}_0$. We also assume that @ $z = \infty$, $w = 0$, in other words, we take soil layers to such depths underneath the pile, where the vertical displacements can be assumed to be negligible.

Response to Harmonic Excitations

By keeping the quantities such as E_p and $E_i, i = 1,..n$ as complex material property functions, one could introduce damping into the vibration system. Assuming that the applied force at the top of the pile is harmonic, i.e.,

$$\widetilde{P}_0 = Pe^{i\Omega t}, \tag{15}$$

then we could assume that the response of the pile is also harmonic, i.e.,

$$\{W\} = \{\overline{W}\} e^{i\Omega t}. \tag{16}$$

Substituting the above equation into the field equation for the pile and simplifying, we get the first component of the force vector as,

$$F_1 = P - \widetilde{M}_0 \Omega^2 = 0. \tag{17}$$

Computational Model

In order to incorporate damping properties in the pile and the soil, the material properties such as Young's modulus of elasticity are considered as complex variables. This requires that the nodal displacements W_i and $\phi(r)$ also have to be complex. The second order differential equation shown in Eq.(13) is converted to a tridiagonal matrix equation using the classical finite difference technique, and the coefficients are complex numbers. In this model, since Eqs. (11) and (13) are coupled, they have to be solved separately in an iterative manner. The values of k and $2t$ which are also complex, are not known initially and therefore some approximate values of k and $2t$ are assumed as equal to modulus of the soil in the beginning. Iterative procedure developed by Vallabhan and Das (1988,1989,1991a, 1991b) and Vallabhan et al. (1991c) for solving beams and plates on elastic foundations is used here. The parameter used for convergence characteristics is γ given by Eq. (14). The iterative procedure is stopped when the difference between two consecutive values of γ are within 0.0001. Several problems were studied to find the number of divisions required for the pile and the soil domain in order to obtain good results. However for the ϕ function, it is recommended to use a distance about 100 diameters away from the pile interface with about 1000 to 2000 number of divisions to obtain accurate slopes of the function especially near the soil-pile interface . In other words, the discretization near the pile interface is quite critical in the analysis. Also, the value of the tolerance need not be that small, even a value equal to 0.001 yields satisfactory results.

An Example Problem

The example problem is that of a pile made of concrete and the analysis starts assuming the pile is end bearing. Richart et. al (1970) have given solutions to frequencies of end bearing piles where the displacement of the pile at the bottom is assumed as zero and the surrounding soil has no strength or any effect on the response of the pile. The problem is solved for this ideal condition first and then some properties of the surrounding soil and the soil at the bottom are introduced to determine their effects on the soil-pile response.

Problem No. 1

The dimensions of the pile and the soil characteristics are given below:

Length of the pile, l = 25 m.
Radius of the pile, R = 200 mm.
Modulus of Elasticity of the pile, E_p = 20 GPa.
Mass density of the pile, ρ_p = 2.4 T/m^3
Maximum distance where ϕ- function is
considered as zero = 40 m.

Case a.
Several values of the mass at the top such as 0, 1000, 2500, 10000 and 25000 Kg. are used at first where all the properties of the soil and damping are assumed to be zero. The values of the resonant frequencies obtained from this model are given in Table 1. Calculations are made with a harmonic force $\tilde{P}_0$ equal to 2000 kN. applied on the top of the pile These values obtained are exactly the same as those calculated by formula given by Richart (1970).

Case b.
Here some properties of the soil such as mass density equal to 1.6 T/m^3 with almost negligible Modulus of elasticity are introduced. The respective values

of the resonance frequencies are respectively shown in Table 1. In other words, even though the soil mass have low strength, when the soil and the pile have perfect compatibility at the soil-pile interface. A very small amount of soil mass vibrate in phase with the pile which reduces the magnitudes of the frequencies slightly. The effect of this coupling is automatically accomplished in this model. However, without the strength of the soil, the difference in the frequencies are relatively small.

Case c.

In this case, the modulus of elasticity of the soil is kept as equal to 20 MPa, but the density of the soil is kept as zero. One can see the sudden influence of the strength of the soil on the resonant frequencies.

Case d.

In this case, the density and modulus of elasticity of the soil are used as 1.6 T/m^3 and 20 MPa. Respectively. It can be seen that there is substantial difference between the values of resonant frequencies in case d with results from cases a, b and c.

Case e

For this case, the value of the modulus of elasticity of the soil is continued below the bottom of the pile for a depth of 37.5 m. At a depth of 37.5 m. below the ground level, there is a very hard stratum. In other words, the vertical displacement is zero at that elevation. The values of the resonant frequencies are shown in Table 1. All these examples illustrate the influence of these additional soil parameters on the dynamic response of the pile and the potentiality of the method to consider these parameters easily.

All the above results presented here are based on a ϕ- function whose value becomes zero at a specified distance r_{max} from the pile. For all solutions presented here, the value of r_{max} is specified as 40 m. For smaller values of r_{max}, the resonant frequencies becomes slightly larger. A graded discretization is desirable to get more accurate results with less number of divisions. More research is necessary to present guidelines for the value of r_{max} and the number of divisions used for the numerical solution of the ϕ- function. Interestingly, even with 2000 divisions in the r-direction, the results are obtained in a fraction of a second on the PC.

Problem 2.

This is the case where the properties of the soil layers are introduced. The details of the pile, the soil layers and soil layers below the pile are shown in Fig. 3. The distance r_{max} is kept as equal to 100 times the diameter. The results of the analysis performed are without any damping in the pile or soils. The dimensions of the pile and the soil characteristics are given below:

Length of the pile, l	= 20 m.
Radius of the pile, R	= 250 mm.
Modulus of Elasticity of the pile, E_p	= 20 GPa.
Mass density of the pile, ρ_p	= 2.4 T/m³
Maximum distance where ϕ- function is considered as zero	= 50 m.
Density of the soil (kept as a constant, but in theory, it can be different) ρ_s	= 1.4 T/m³.

.

The resonant frequencies for values of various masses are computed and shown in Table 2.

Conclusions

A simple method to compute the axial vibration response of piles placed in a multi-layered soil medium is given here. Since the horizontal component of displacement of the soil is small and it is ignored in this model. The method is consistent with the principles of solid mechanics and assumes perfect compatibility of displacements with in the layers of the soil medium. There is substantial difference between the values of frequencies computed using Richart's formula and those made incorporating the values of the mass density of the soil and modulus of elasticity of the soil. This is because a certain mass of soil is vibrating in phase with the pile, in addition to the shear resistance of the soil on the sides of the pile. Here, we assume perfect compatibility of displacement between the soil and the pile. Also, when the stiffness of the soil at the bottom of the pile is introduced instead of a perfectly rigid base, the frequencies reduced slightly. More research is necessary to improve the accuracy of the model comparing these results with more sophisticated finite element models.

Acknowledgment
The author wishes to acknowledge Mr. Mukkamala A. Sarma who helped to prepare the drawings.

Appendix I. Nomenclature

E_p, A_p, R, l	Modulus of elasticity, Area of cross section, radius and length of the pile.
E_i, μ_i	Modulus of elasticity and Poisson's ratio of soil in region i = 1,2.
$k_i, 2t_i$	Soil parameters in the i-layer
m, n	Coefficients in the differential equation defining ϕ.
r, z	Cylindrical coordinates defining the problem.
$w(z,t)$	Axial displacement of the pile.

$\overline{w}(r, z, t)$ Displacement of the soil in the z- direction.

$\alpha, \beta, \gamma, \varepsilon$ Parameters used to obtain a consistent solution.

r_{max} The distance from the pile interface to a point where the ϕ- function is negligible.

$\sigma_{ij}, \varepsilon_{ij}$ Stress and strain tensors on the soil.

$\tilde{P}_0, \tilde{M}_o$ Magnitude of Applied force and Mass on top of the pile.

Appendix II. References

Kuhlemeyer, R. (1978), "Vertical Vibration of Piles", Journal of Geotechnical Engineering, ASCE, Vol.105, No. GT2, February, pages (273-287)

Novak, M. (1977), "Vertical Vibration of Floating Piles", Journal of Engineering Mechanics, ASCE, Vol.103, No. EM1, February, pages (153-167)

Nogami, T. and Novak, M., (1976), "Soil-Pile Interaction in Vertical Vibration", Earthquake Engineering and Structural Dynamics, Vol. 4, pages (277-293).

Nogami, T., Zhu, J.X. and Itoh, T., (1992), "First and Second order Dynamic Subgrade Models for Soil-Pile Interaction Analysis", ASCE STP on Piles under Dynamic Loads, GSP No. 34, Geotechnical Engineering Div., ASCE, Sept., pp. 187-206.

Prakash, S., (1992), "Piles Under Dynamic Loads", Geotechncial Special Publication No. 34, ACSE, September.

Baranov, V.A., (1967), "On the Calculation of an Embedded Foundation" (in Russian), Voprosy Dinamiki I Prochnosti, No. 14, Polytechnic Institute of Riga, pages (195-209).

Vallabhan, C.V. Girija., (1995), "Settlement of Axially Loaded Pile Foundations", paper presented to Texas ASCE, Lubbock, Texas in 1994, also accepted for publication in the International Journal for Engineering Analysis and Design, Edited by J.N. Reddy .

Vallabhan, C.V.Girija. and Mustafa, G., (1995), "A New Model for the Settlement Analysis of Drilled Piers", to be published, International Journal of Analytical and Numerical Methods in Geomechanics", edited by C.S. Desai.

Seed, H. B. and Reese, L. C. (1957). "Action of Soft Clay Along Friction Piles." Trans. ASCE. Vol. 122, 731-754.

Vallabhan, C. V. G. and Das, Y. C. (1988). "A parametric study of beams on elastic foundations." *J. Engrg. Mech.,* ASCE, 2072-2082.

Vallabhan, C.V.G., and Das, Y.C., (1991a)," A Refined Model for Beams on Elastic Foundations", *Int. J. Solids and Structures,* Vol. 27,5, 629-637.

Vallabhan, C.V.G., and Das, Y.C., (1991b), "Modified Vlasov Model for Beams on Elastic Foundations", *J. Geotech. Engrg. ,*ASCE, Vol.117, 6, 956-966.

Vallabhan, C.V.G., Straughan, W.T. and Das, Y.C., (1991c), " Refined Model for Analysis of Plates on Elastic Foundations", *J. Engrg. Mech.,* Vol. 117,12, 2830-2844.

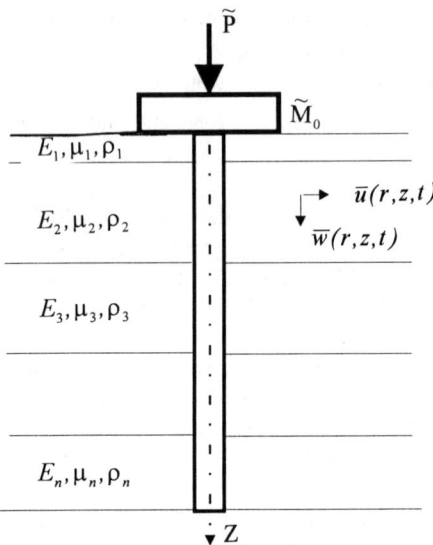

Fig. 1. Pile in a Multi Layered Soil Medium

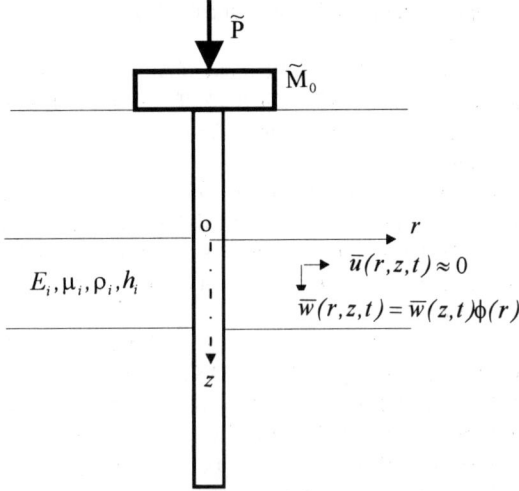

Fig. 2. A Strip of Pile and Soil in Local Coordinates

Table. 1

Mass (kg)	Case a $\rho_s=0$ $E_1=0$ $E_{bottom}=\infty$	Case b $\rho_s=1.6T/m^3$ $E_1=0$ $E_{bottom}=\infty$	Case c $\rho_s=0$ $E_1=20\ Mpa$ $E_{bottom}=\infty$	Case d $\rho_s=1.6T/m^3$ $E_1=20\ MPa$ $E_{bottom}=\infty$	Case e $\rho_s=1.6T/m^3$ $E_1=20\ MPa$ $E_{bottom}=20\ MPa$
		Resonant Frequency (rad/sec)			
0	181.25	178.6	270.5	304.3	269.7
1000	160.25	158.5	237.2	264.2	243.4
2500	137.75	135.8	201.6	220.7	208.5
10000	89.25	88.70	126.2	132.6	128.5
25000	60.25	60.40	84.7	86.40	84.50

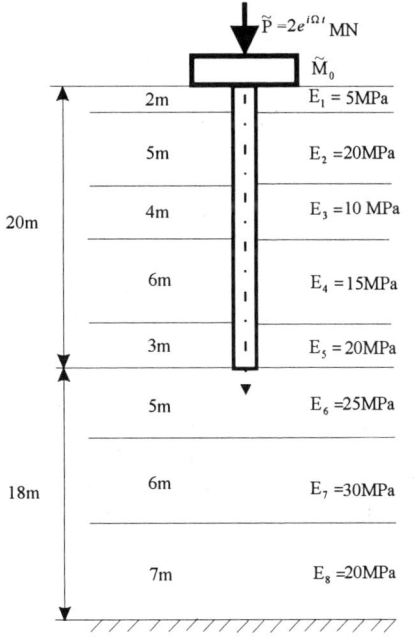

Table. 2.
Results of Problem No. 2

Mass $\widetilde{M}_0$ (Kg)	Resonant Frequency (rad/sec)
0	214.7
1000	200.6
2500	181.5
10000	126.5
25000	87.5

Fig.3. Pile in a Layered Soil Medium

Centrifuge and Numerical Modeling of Soil-Pile Interaction During Earthquake Induced Soil Liquefaction and Lateral Spreading

T. Abdoun[1], R. Dobry[2], and T. D. O'Rourke[3]

Abstract

A cooperative project is reported between teams at Rensselaer Polytechnic Institute (RPI) and Cornell University to investigate the effects of lateral spreading on pile foundations. Centrifuge tests of lateral spreading and the corresponding permanent bending moments measured in instrumented model piles conducted at RPI, are used to study the problem and verify and calibrate Cornell University's computer program B-STRUCT. Individual piles and pile groups are investigated in a variety of pile and soil configurations. The paper reports in detail the results of Model 1, corresponding to the simulation of reinforced concrete piles that were deformed by lateral spread under the NFCH building in Niigata during the 1964 earthquake.

Introduction

Lateral spread of liquefied soil in sloping ground or near a waterfront is a cause of significant damage to pile foundations of buildings and port facilities. Cases of distress of deep foundations due to this reason have been reported in a number of seismic events in Japan, the U.S. and other countries, including: Niigata 1964, Loma Prieta 1989, and Hyogoken-Nambu 1995 earthquakes (Hamada et al., 1986; Benuzka, 1990; Tokimatsu et al., 1996). The volumes of case histories of liquefaction and ground deformation in Japanese and U.S. earthquakes published by Hamada and O'Rourke (1992) and O'Rourke and Hamada (1992) include descriptions of several of these and other cases of pile foundation distress associated with lateral spreads. Figure 1 sketches typical deep foundation damage observed near the Kobe waterfront in the 1995 Hyogoken-Nambu earthquake (Tokimatsu et al., 1996).

---

[1]Graduate Research Assistant, Dept. of Civil Engineering, Rensselaer Polytechnic Institute, Troy, NY

[2]Professor, Dept. of Civil Engineering, Rensselaer Polytechnic Institute, Troy, NY

[3]Professor, School of Civil & Environmental Engineering, Cornell University, Ithaca, NY

In the last 5-10 years, the study of this problem, including development of evaluation methods, has focused on well-documented case histories of reinforced concrete piles damaged in the 1964 Niigata earthquake (Hamada et al., 1986; Miura and O'Rourke, 1991). Program B-STRUCT was developed for this purpose at Cornell University. This is a beam-on-elastic foundation code including nonlinear soil springs and a nonlinear moment-curvature relation for the pile, where the lateral spreading effect is modeled by displacing laterally the soil spring supports by an amount equal to the free field permanent deformation (Meyersohn, 1994). Simultaneously, a technique to model realistically liquefaction and lateral spread of a sloping saturated sand deposit, using an inclined laminar box excited by base shaking was developed at Rensselaer Polytechnic Institute, RPI (Taboada, 1995; Dobry et al., 1995). In another centrifuge project, the effect of liquefaction on the nonlinear p-y curves of saturated sand was studied at RPI by in-flight loading tests of centrifuge model piles instrumented with strain gages (Liu and Dobry, 1995; Dobry et al, 1995). In the cooperative RPI-Cornell project reported in this paper, these three efforts are further developed and applied to the study of pile response in the presence of lateral spreading. Specifically, the development of pile displacements and bending moments due to lateral spread of layered deposits in the free field is modeled in the centrifuge, and the results are used to verify and calibrate program B-STRUCT for both individual piles and pile groups.

Pile 2 under the Niigata Family Court House (NFCH) and its response in 1964 (Figure 2) was selected for initial study and is reported in this paper as centrifuge Model 1. The 1964 NFCH building is a well documented case history for which information regarding pile properties, subsurface conditions, and extent of ground displacement are available (Hamada, et al., 1986; Kawashima, et al., 1988). The NFCH building was founded on 35 cm diameter concrete piles (Figure 2). The measured offset between the two ends of the pile after the earthquake was approximately 70 cm for Pile 2, which penetrated about 1 m into the lower nonliquefied soil layer. Figure 3 presents the pile bending moments for several values of lateral spreading, predicted with program B-STRUCT before to the centrifuge testing was done (Meyersohn, 1994).

Determination of p-y curves in the centrifuge

In American practice, the nonlinear lateral soil spring at a given depth is typically specified as a p-y curve, where p is the soil resistance per unit length of pile, and y is pile lateral deflection. These p-y curves for various soils have been empirically backfigured from field tests where an instrumented pile is laterally loaded at the head (Reese and Wang, 1993). The same approach was used at RPI to obtain the p-y curves, by laterally loading in-flight an instrumented centrifuge pile model embedded in the soil. The technology and the setup used for this determination of p-y curves in the centrifuge are the same reported by Liu and Dobry (1995) and Dobry et al. (1995).

Computer code LPILE (Reese and Wang, 1993) was used for analyzing the test data. The pile-soil models used for p-y curve determination were accommodated in a rigid rectangular container 53.34 cm (length) by 20.54 cm (width) by 20.32 cm (height). Figure 4 shows a typical profile of the pile-soil model. The soil surrounding the pile is 18 cm high, which at 40g centrifugal acceleration corresponds to 17.2 m in prototype.

The model pile was a 0.95 cm diameter brass tube of 0.036 cm wall thickness. This model simulates a prototype pipe pile of outside diameter, d = 38 cm, with a bending stiffness, EI = 28,500 $kN - m^2$. Both ends of the pile were fixed, with the tip fixed to the container, while the pile head is fixed without rotation to a lateral loading unit (Figure 4). The pile head lateral force was monitored by a load cell, while its lateral displacement was measured by an LVDT. Five pairs of full-bridge circuited strain gages were installed along the surface of the model pile to monitor the bending moments. The lateral loading consisted of a slow sinusoidal signal of 5 cm prototype amplitude and loading period of 40 sec, also in prototype units.

p-y curves

Cemented sand: As discussed in next section, a cemented sand layer was used to model the nonliquefiable sand in the field during the lateral spreading tests. The centrifuge experiment of Figure 4 was used to define the set of p-y curves which best represents the cemented sand-model pile system. The measured lateral load and displacement at the top of the pile were used for first estimates of the p-y curves along the pile, using computer code LPILE. This step was then repeated for several iterations until the calculated moments using LPILE and the recorded moments along the pile matched. The final p-y curves constructed this way are compared in Figure 5 with the p-y curves for dense sand recommended by Reese and Wang (1993), for various values of the depth/pile diameter ratio, z/d. The calculated bending moment profile using LPILE in conjunction with this set of cemented sand p-y curves is compared with the in-flight strain gage measurements in Figure 6.

Nevada sand, $D_r = 40\%$: The p-y curves of the saturated Nevada sand layer ($D_r = 40\%$) before liquefaction, were obtained in the centrifuge using the same setup of Figure 4 and computer code LPILE. The p-y curves obtained this way are plotted in Figure 7. The calculated bending moment profile using this set of p-y curves and LPILE is compared with the recorded moments in Figure 8. The next step is to define the p-y curves of this saturated sand after liquefaction. The soil lateral resistance of fully liquefied soil can be expressed as $p_d = C_u p$, where p_d is the degraded soil lateral resistance after liquefaction, p is the resistance before liquefaction for the same displacement, and C_u is a degradation coefficient. For fully liquefied Nevada sand of $D_r = 60\%$, $C_u \approx 0.1$ (Liu and Dobry, 1994). It was assumed that $C_u = 0.1$ could be used for liquefied Nevada sand of $D_r = 40\%$; the corresponding p-y curves are also plotted in Figure 7.

Laminar box and model pile

A laminar box is used for the centrifuge modeling of the 1964 Niigata earthquake NFCH building case history. The container inside dimensions are 45.72 cm (length) by 25.40 (width) by 26.39 cm (height),see Figure 9. The box consists of a stack of up to 39 rectangular rings separated by linear roller bearings, arranged to permit relative movement between rings in the long direction with minimal friction (Taboada, 1995; Dobry et al., 1995).

The model pile used in the laminar box model setup was made of 0.95 cm in diameter polyetherimide rod (ULTEM 1000). At 50g centrifugal acceleration, this model simulates a prototype pile of diameter d=47.5 cm and bending stiffness, EI = 8,000 $kN - m^2$. This value of EI is within the range of effective stiffnesses of the NFCH reinforced concrete piles in 1964, which ranged from an initial EI= 18,000 $kN - m^2$ to an effective secant EI= 4,500 $kN - m^2$ after cracking of the concrete pile took place. The model pile is quite strong and remained elastic during the tests. Because no yielding occurred during testing the same model pile could be used in several experiments.

Six pairs of full-bridge circuited strain gages were installed along the surface of the model pile to monitor bending moments during lateral spreading. Two pairs of strain gages were placed near each interface (Figure 9). As large deformations were expected in the test, microcrystalline wax and a soft plastic shrink tube were used to waterproof the strain gages. Sand grains were glued to the shrink tube surface to develop an adequate pile-soil roughness.

Centrifuge modeling of Pile 2 in NFCH building

Model 1, simulating approximately Pile 2 of the NFCH building, involved a single pile (EI=8000 $kN - m^2$) embedded in a three-layer system (Figure 9). This graph shows the soil profile in model units as well as the instrumentation used. The test was done at 50g centrifugal acceleration. The total height of the profile is 20 cm in model units, that is 10 m in prototype units. In prototype units, the top layer is a 2 m cemented sand with a cohesion of 6.5 kg/cm^2, which models the nonliquefiable sand in the field, followed by a 6 m layer of liquefiable uniform Nevada sand placed at a relative density of about 40%, followed by a 2 m layer of the same cemented sand. The soil profile is fully saturated with water and the model is inclined 2° to the horizontal. Both top and bottom cemented sand layers were perforated to make them pervious and avoid accumulation of water at the bottom of the top layer. The use of cemented sand to model the top non-cemented, unsaturated sand layer in the field, followed several unsuccessful attempts to use non-cemented sand in the model. In what follows, prototype units are consistently used.

The input acceleration time history applied at the base of the soil profile is shown in Figure 10. Accelerations and excess pore pressure ratios recorded in the soil during the test are shown in Figures 10 and 11, respectively. The recorded accelerations and excess pore pressure ratios indicate that the 6 m, 40 % relative density Nevada sand layer did liquefy, while as expected the cemented sand remained solid. The profiles of soil lateral displacement measured by the LVDT's mounted on the laminar box rings at different times during shaking, and at the end of shaking, are shown in Figure 12. The soil lateral deformation at the top after shaking is about 80 cm. All soil accelerations, pore pressures and lateral displacements measured in this test with a model pile were very similar to those measured in a preliminary experiment without a pile, revealing that the presence of the pile did not affect the free field response and lateral spreading of the soil. Figure 13 shows a comparison between the pile displacement measured at the top of the soil profile at the end of shaking and the pile displacement calculated using the program B-STRUCT. It is interesting that the measured tip pile displacement at the top of the pile is larger than the free field soil displacement; the reason for this is discussed below.

Figure 14 shows the bending moment time histories recorded at different depths along the pile during shaking. The moments recorded within the top 4 m kept increasing with time until local soil failure occurred around the pile in the top cemented sand layer. This soil failure was verified by direct observation during excavation of the soil around the pile after the test. Due to this soil failure, which was also responsible for the larger pile displacement relative to the soil in Figure 13, the force applied by the cemented soil on the pile decreased, causing the recorded decrease in bending moments. Figure 15 presents the profiles of recorded pile bending moments at different times during shaking. Figure 16 shows that the bending moments recorded in the centrifuge model pile compare well with the Cornell University analytical results for Pile 2 at the NFCH building during the 1964 Niigata earthquake, for the same soil surface lateral displacements in the free field, D_H, up to $D_H = 24$ cm.

Conclusions

The previous results for centrifuge Model 1 indicate that:

1) The presence of the pile did not affect the free field response of the soil, as revealed by a comparison between these results and those of the free field test with no pile. Especially important is the consistency between lateral surface soil displacement in the tests with and without pile: about 80 cm in both cases (Figure 13). This displacement is also comparable to that measured after shaking at the NFCH building site in the 1964 Niigata earthquake.

2) In Model 1, the cemented soil in the top layer failed around the pile when the bending moments reached about 150 to 175 $kN - m$, with corresponding decrease in bending moments afterwards (Figures 14 and 15), as well as pile penetration into the soil (Figure 13). This soil failure and pile penetration were well predicted by Cornell University program B-STRUCT, indicating that the p-y curves measured at RPI for the different layers and the Cornell University analytical technique correctly account for this important aspect of the response.

3) Up to a surface ground displacement of about 24 cm (Figure 16) centrifuge Model 1 measurements are very consistent with the analytical results for the NFCH concrete pile in Niigata in 1964 using the program B-STRUCT. This is the range during which both soil and pile exhibited an approximately linear behavior, both in the field and in the centrifuge. Soil failed in the top layer around the elastic pile model in the centrifuge test when the bending moment reached about 150 to 175 $kN - m$, while the concrete pile under the NFCH building exhibited strong nonlinear moment-rotation response due to concrete cracking when the bending moments reached about 90 $kN - m$. The effects in both the centrifuge and the field tended to reduce the pile moments in the top layer, but for different reasons. If the concrete pile in Niigata had remained linear at these high moments, the soil near the pile would have failed as it happened in the centrifuge model test.

4) In the middle liquefied layer, the free field soil displacement varies linearly with depth (Figure 12) and the pile moments also vary linearly (Figures 14-16). This last observation is as predicted by beam theory for soil having no resistance, and is consistent with the very large degradation ($C_u \approx 0.1$) of the p-y curves used in the calculations for the liquefied soil.

Acknowledgment

The authors want to thank Drs. Korhan Adalier, Lu Liu and Victor Taboada for their help and advice toward the centrifuge model tests. This research was supported by the National Center for Earthquake Engineering Research (NCEER), Buffalo, NY.

Reference

Benuzka, L. Ed (1990), "Loma Prieta earthquake reconnaissance report." report by EERI and NRC, Supplement to Vol. 6 of Earthquake Spectra.

Dobry, R., V. Taboada, and L. Liu (1995), "Centrifuge modeling of liquefaction effects during earthquakes." Proc. First Intl. Conf. on Earthquake Geotechnical Engineering, Tokyo, Japan.

Hamada, M., and T. D. O'Rourke, (Eds., 1992), "Case studies of liquefaction and lifeline performance during past earthquakes, Vol. 1: Japanese case studies." NCEER, SUNY-Buffalo, NY.(Tech. Rept. NCEER-92-0001)

Hamada, M., S. Yasuda, R. Isoyama, and K. Emoto (1986), "Study on liquefaction-induced permanent ground displacements." Association for the Development of Earthquake Prediction, Tokyo, Japan.

Kawashima, K., K. Shimizu, S. Mori, M. Takagi, N. Suzuki, and D. Nakamura (1988), "Analytical studies on damage to bridges and foundation piles caused by liquefaction-induced permanent ground displacement." Proceedings, 1st Japan-U.S. Workshop on Liquefaction, Large Ground Deformations and Their Effects on Lifeline Facilities, Tokyo, Japan, pp. 99-117.

Meyersohn, W. D. (1994), "Pile response to liquefaction-induced lateral spread." Ph.D. Thesis, Cornel University, Ithaca, NY.

Miura, F. and T. D. O'Rourke (1991), " Nonlinear analysis of piles subjected to liquefaction-induced large ground deformation." Proc. 3rd Japan-U.S. Workshop on Earthquake-Resistant Design of Lifeline Facilities and Countermeasures for Soil Liquefaction, NCEER, SUNY-Buffalo, NY. (Tech. Rept. NCEER 91-0001)

Liu L. and R. Dobry (1995). "Effect of liquefaction on lateral response of piles by centrifuge model tests." NCEER Bulletin, SUNY-Buffalo, NY., pp. 7-11.

O'Rourke, T. D., and M. Hamada, (Eds., 1992), "Case studies of liquefaction and lifeline performance during past earthquakes, Vol. 2: United States case studies." NCEER, SUNY-Buffalo, NY. (Tech. Rept. NCEER-92-0002)

Reese, L. C. and S. T. Wang (1993), "Documentation of computer program LPILE version 4.0." Ensoft, Inc., Austin, Texas.

Taboada, V. (1995), "Centrifuge modeling of earthquake-induced lateral spreading in sand using a laminar box." Ph.D. thesis, Rensselaer Polytechnic Institute, Troy, NY.

Tokimatsu, K., H. Mizuno, and M. Kakurai (1996), "Building damage associated with geotechnical problems." Soil and Foundations, Special Issue of Geotechnical Aspects of the 1995 Hyogoken Nanbu Earthquake, January, pp. 219-234.

Yoshida, N. and M. Hamada (1991), "Damage to foundation piles and deformation pattern of ground due to liquefaction-induced permanent ground deformations." 3rd Japan-U.S. Workshop on Earthquake Resistant Design of Lifeline Facilities and Countermeasures for Soil Liquefaction, Technical Report NCEER 91-0001, NCEER, Buffalo, NY, pp. 147-161

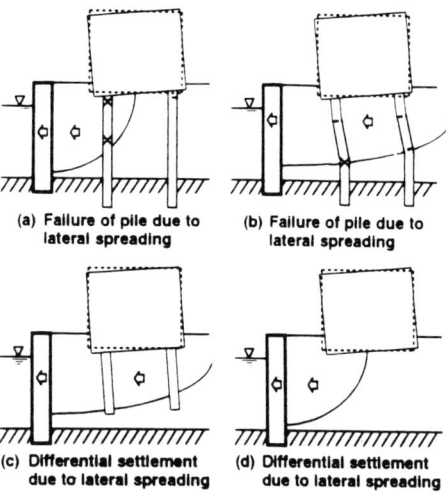

Figure 1: Typical foundation damage due to lateral spreading in 1995 Kobe earthquake (Tokimatsu et al., 1996)

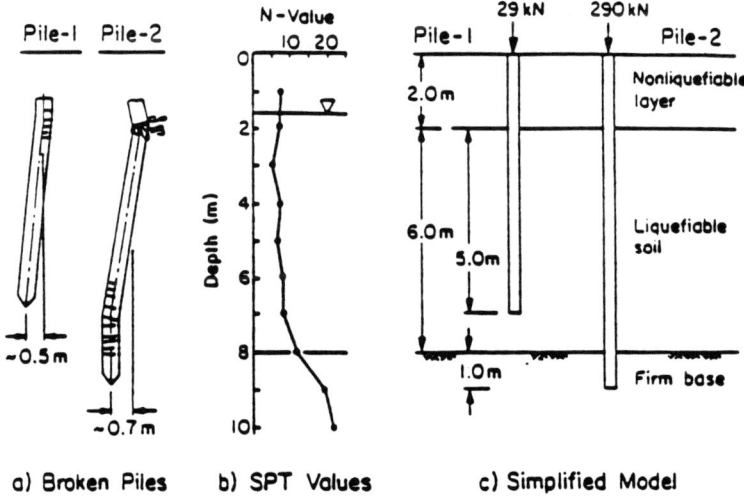

Figure 2: Observed pile deformation at NFCH building, Niigata earthquake (Yoshida and Hamada, 1991; Meyersohn, 1994)

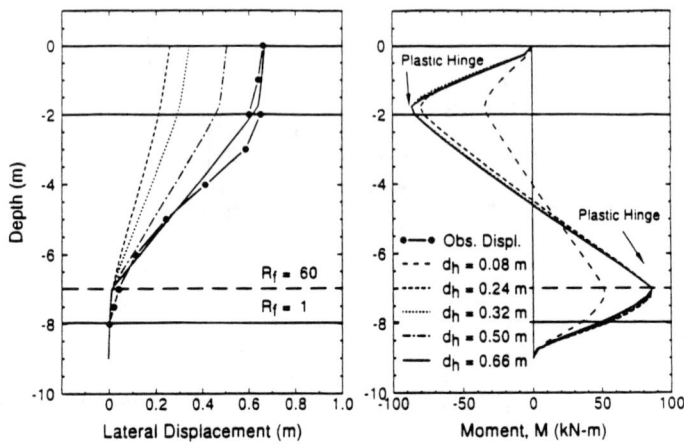

Figure 3: Analytical results using program B-STRUCT for Pile-2 at NFCH building (Meyersohn, 1994)

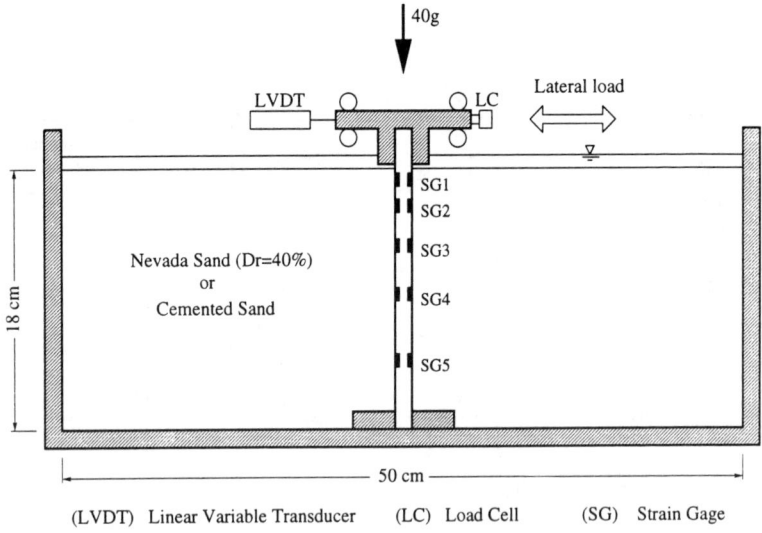

Figure 4: Schematic of the pile-soil centrifuge model used to determain the p-y curves, modified from Liu and Dobry, 1995

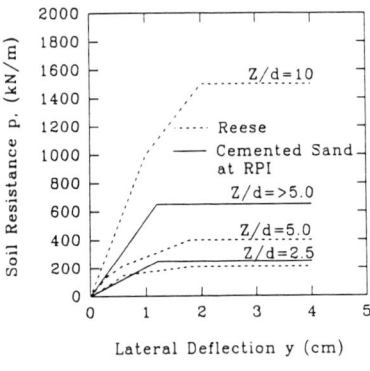

Figure 5: p—y curves for the cemented sand from centrifuge test of Figure 4 and Reese p—y curves for dense sand

Figure 6: Measured and calculated moments for the cemented sand test of Figure 4

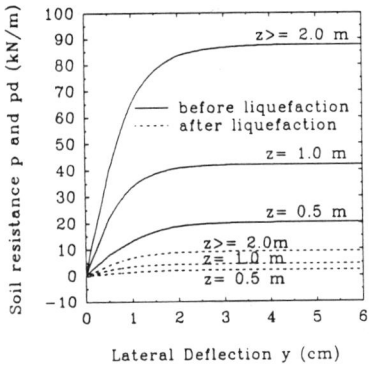

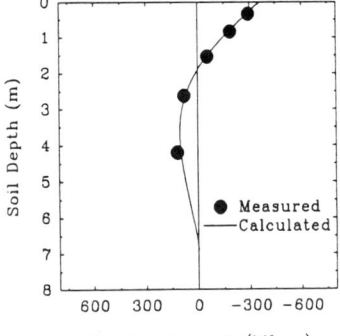

Figure 7: p—y curves for the Nevada sand (Dr=40%) from centrifuge test of Figure 4

Figure 8: Measured and calculated moment for the Nevada sand (Dr=40%) test of Figure 4

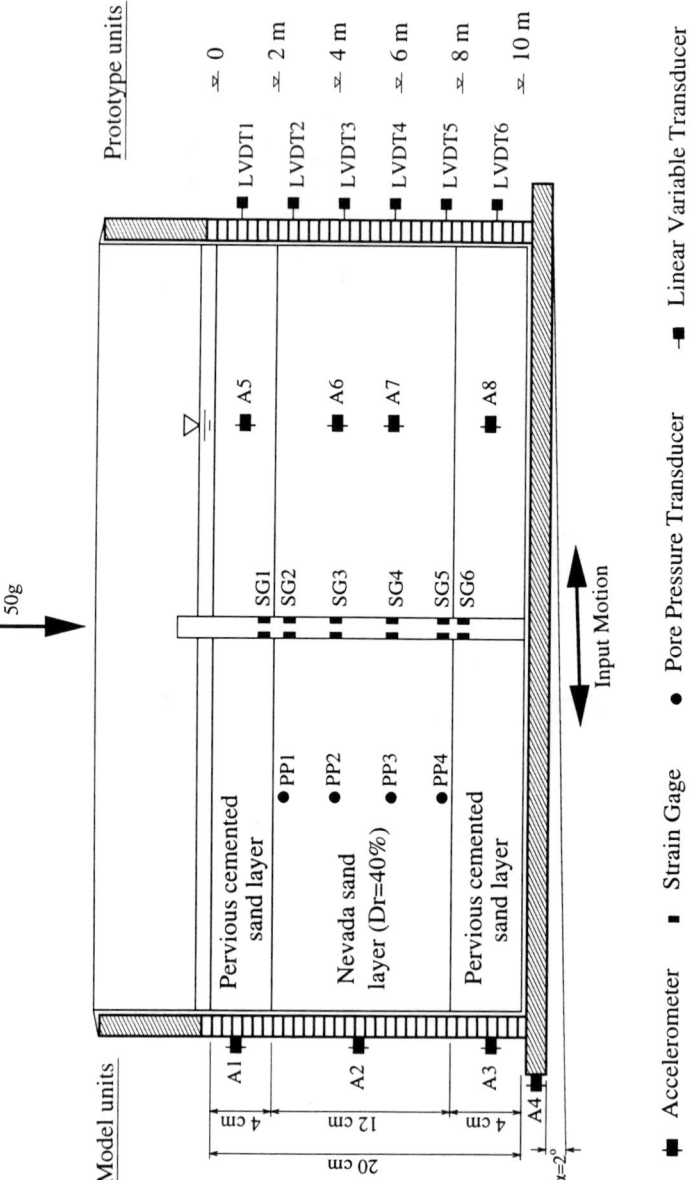

Figure 9: Centrifuge lateral spreading model setup of Pile-2 at NFCH building, using RPI laminar box

■ Accelerometer   ■ Strain Gage   ● Pore Pressure Transducer   ┤■ Linear Variable Transducer

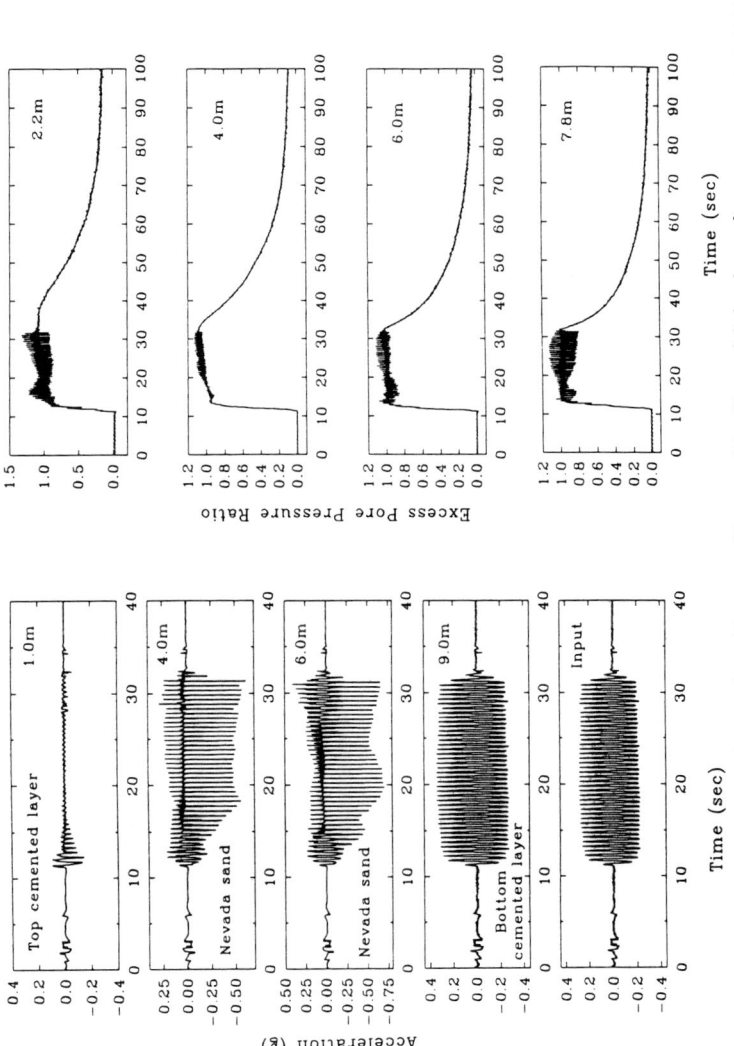

Figure 11: Time histories of excess pore pressure ratio in the Nevada sand layer (Dr=40%)

Figure 10: Time histories of accelerations recorded in the soil during shaking

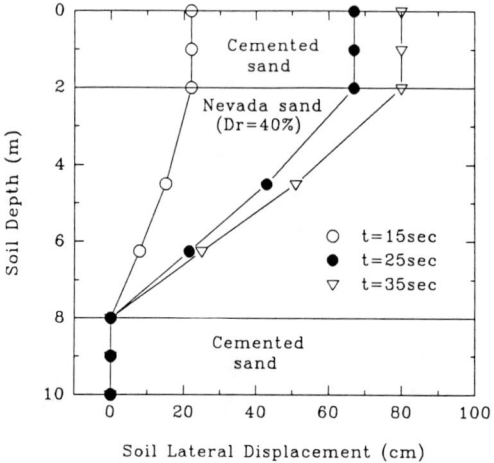

Figure 12: Lateral displacement soil profiles measured during shaking

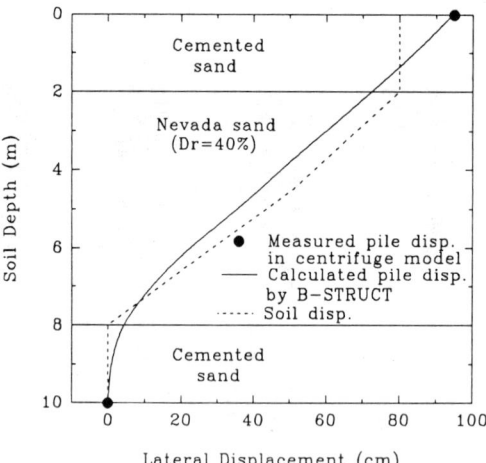

Figure 13: Soil and pile lateral displacement profiles at end of shaking

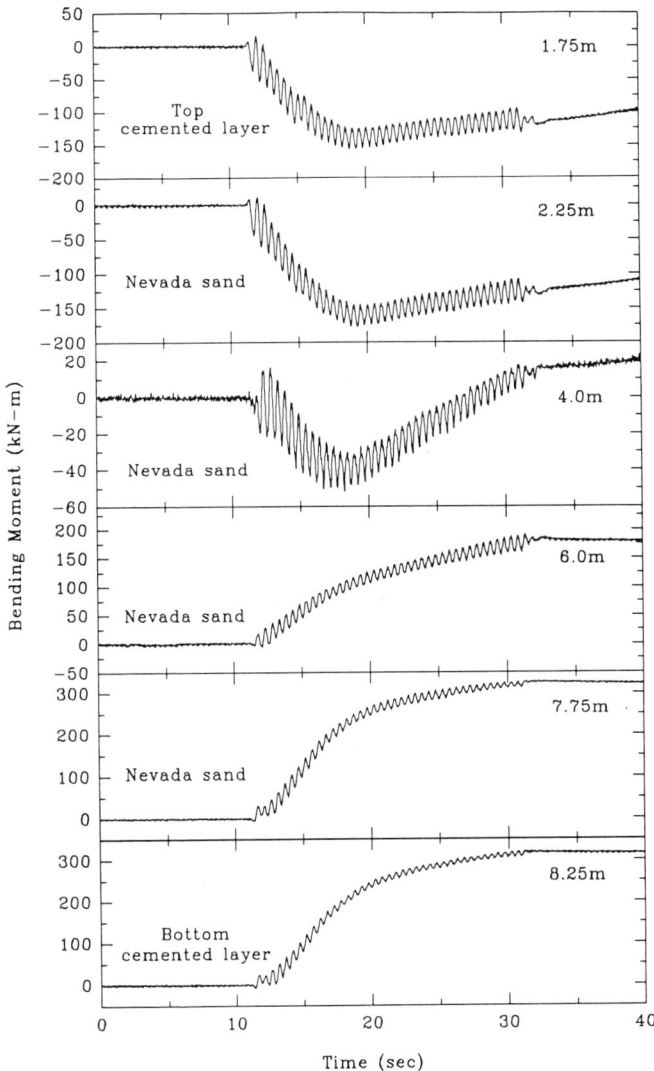

Figure 14: Time histories of measured bending
moments along the pile

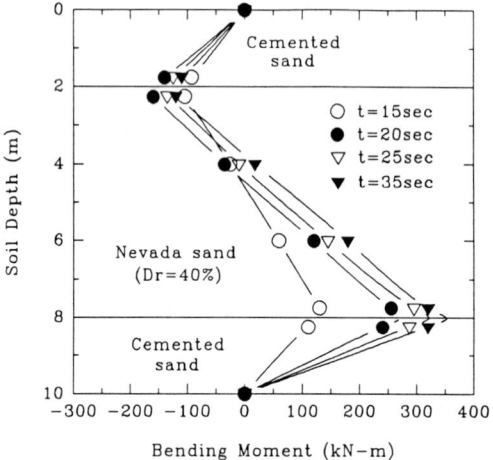

Figure 15: Measured bending moments profiles during shaking

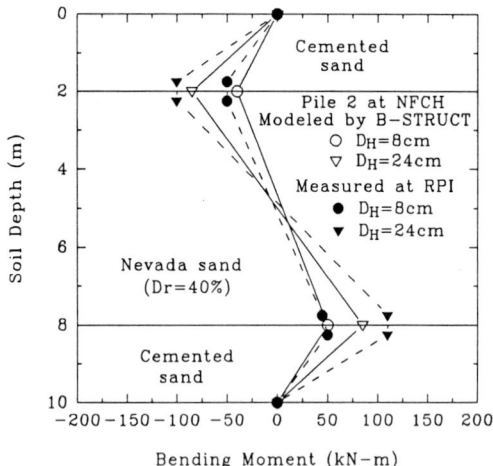

Figure 16: Comparison between bending moments measured in RPI centrifuge test and those calculated by B-STRUCT for Pile 2 at NFCH building in the 1964 Niigata earthquake

Simulation of Soil-Structure Interaction on a Shaking Table

Kazuo KONAGAI[1] and Toyoaki NOGAMI[2]

Introduction

Model experiments on a shaking table are useful for the identification of important phenomena and verification of predictive theories regarding the dynamic behavior of a prototype structure during an earthquake; a shaking table is, in general, controlled so that it closely follows the input *free-field* motion. In reality, however, a structure interacts with its foundation on or in the ground, and responds differently. This interaction causes the motion of the ground at the structure's base to deviate from the free-field motion. This effect may be partly incorporated into an experiment by filling a bin on the table with the actual prototype soil and then putting the model on it. This method is particularly useful when non-linear features of the soil in the vicinity of a structure must be considered. But the process of preparing a soil model is difficult; and even when prepared, it still can not allow for the effect of wave-dissipation into an infinite soil medium that exists in the field.

Shaking tables of many different sizes have been used so far. Some are quite large, allowing models with dimensions of several meters to be shaken. However, even these tables are not always large enough for all structural models of interest to be tested. Within the finite base size of a shaking table and within the limit of its dynamic loading capacity, only one part of a whole structure, such as devices for vibration reduction, can be tested. In this case also, the input motion to the model's base will be affected by the presence of the model.

This paper introduces one idea for controlling a shaking table so that the soil-structure or base-structure interaction effect is incorporated. In order for the interaction effect to be reflected in a shaking table test, a signal equivalent to the further displacement induced by the interaction is added to the input ground or base motion. This method therefore premises a device that can generate signals identical to the transient motion of its base on a soil medium of infinite extent. It is shown

---

[1] Dr. Eng., Assoc. Prof., Inst. of Industrial Science, Univ. of Tokyo, 7-22-1 Roppongi, Minato-ku, Tokyo 106 Japan.
[2] Ph.D., Prof., Civil and Environmental Engineering, Univ. of Cincinnati, Cincinnati, OH 45221, U. S. A.

herein that a variety of unit-impulse response functions of bases or soil mediums overlaid with structures are closely approximated by summing up the basic functions which can be generated by one analog circuit. To all intents and purposes, an analog circuit loses no time in responding to its input signal. This method can therefore be applied in a variety of soil-structure or base-structure interaction experiments without the necessity for physical soil models.

Simulation of Soil-Structure or Base-Structure Interaction in Shaking Table Tests

Shaking tables are usually driven by either servo-hydraulic actuators or electro-magnetic actuators. A model on a shaking table, however, interacts with the table, and often gives the table a force beyond the capacity of its actuators causing the motion of the table to deviate from the intended time history. For this reason, the input signal can be modified so that the motion of the table eventually follows the intended time history of displacement. However, the actual soil, which can be viewed as a natural shaking table, is not stiff enough for the motion at the structure support point to be completely identical to the free-field earthquake motion. When the dominant frequency of a seismic input motion is tuned to the resonance frequency of the structure, for instance, the interaction results in the motion of the soil at the structure's base stopping naturally at this particular frequency.

Thus, in the present method, a shaking table's motion is controlled directly following the actual process of soil-structure interaction. **Figure 1** shows a schematic view of the set-up in a shaking table test for earthquake simulation, in which a superstructure model is placed directly on the table without a physical ground model. The soil-structure interaction effects are simulated by adding appropriate soil-structure interaction motions to the free-field ground motions at the shaking table. In the simulation, first, the transducers at the base of the foundation pick up the signals of the base forces, p_x, p_z and p_θ in sway, vertical and rocking motions, respectively. These three amplified signals are then applied to the circuits h_x, h_z and h_θ to

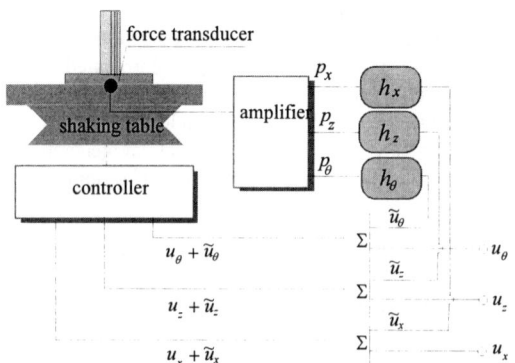

Figure 1 Simulation of soil-structure interaction on a shaking table
(displacement-controlled)

produce the outputs corresponding to the soil-structure interaction motions, $\widetilde{u}_x$, $\widetilde{u}_z$ and $\widetilde{u}_\theta$, respectively. The output signals are then added to the signals of free-field motions, u_x, u_z and u_θ, to produce the signals of foundation motions, $u_x + \widetilde{u}_x$, $u_z + \widetilde{u}_z$ and $u_\theta + \widetilde{u}_\theta$. The signals are translated into the shaking table motions by the shaking table controller.

Actuators used in the above-mentioned system are displacement-controlled, and at the same time, force-controlled actuators are available even though they are not expected to be used in the conventional shaking table systems. If the force-controlled actuators were used, however, base-structure interaction would be reflected in shaking table tests just by reversing the flows of input and output signals (**Figure 2**): Namely, displacement signals from the base of the model, $u_x + \widetilde{u}_x$, $u_z + \widetilde{u}_z$ and $u_\theta + \widetilde{u}_\theta$, are picked up by displacement transducers. Three components of the free-field ground motion, u_x, u_z and u_θ, are then subtracted from the base displacements to obtain signals identical to soil-structure interaction motions, $\widetilde{u}_x$, $\widetilde{u}_z$ and $\widetilde{u}_\theta$. These signals of interaction motions are applied to the circuits S_x, S_z and S_θ to produce transient interaction forces p_x, p_z and p_θ at the model's base. The actuators are, then, so controlled that the output signals from the force-transducers follow closely the interaction forces.

Needless to say, analog circuits h_x, h_z and h_θ (or S_x, S_z and S_θ) are the most important elements in the systems mentioned above, and flexibility (or stiffness) formulation in the time domain is to be discussed in the following chapter to figure out the essential functions of the circuits.

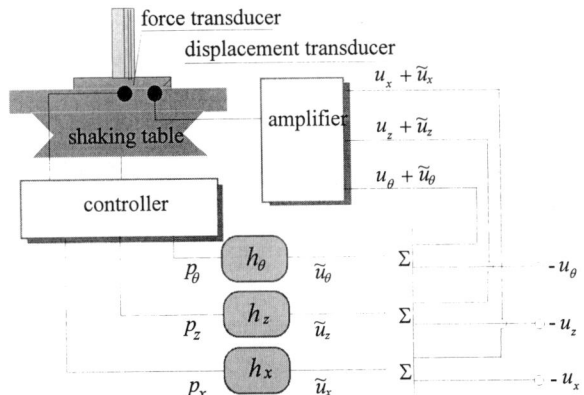

Figure 2   Simulation of soil-structure interaction on a shaking table
(force-controlled)

Flexibility Formulation in the Time Domain

A number of attempts to clarify various aspects of soil-structure interaction have been carried out mostly in the frequency domain, and at the same time, no small number of attempts have been made to provide both rigorous and approximate expressions in the time-domain (See Wolf, 1988). Though there are a variety of soil-structure types to be discussed in the engineering practice, a foundation embedded in a horizontally-layered soil medium (**Figure 3a**) is taken as a representative example. Following some previous numerical examples by Tajimi (1969) and Veletsos (1994) for example, the soil mediums at the side and at the bottom of the foundation are discussed separately, and the soil medium beneath the foundation is assumed to be considerably stiff compared with the soil medium at the side.

Soil-Structure Interaction at the Side
Many wave motion problems of this kind are best expressed in terms of cylindrical coordinates (r, θ, z). Ignoring vertical ground motion of the soil medium at the side of the foundation makes the solution of the ground motion separable (Tajimi (1969), Novak and Nogami (1977)) as:

$$u_r = \sum_m A_m(r, \theta,) \cdot Z_m(z) \tag{1}$$

and thus greatly simplifies the analysis. The contribution of a particular vibration mode $Z_m(z)$ is now discussed. Given the prescribed vibration mode $Z_m(z)$, the surface layer is modeled by a plane of infinite extent supported by Winkler type springs as shown in **Figure 3b**. Lame's constants λ_m, μ_m (μ_m = shear modulus) and mass density ρ_m of the soil plane and Winkler-type spring constant k_m for this vibration mode are obtained by equating all kinds of energies and works, done by forces induced in the interior of the surface soil medium undergoing this particular

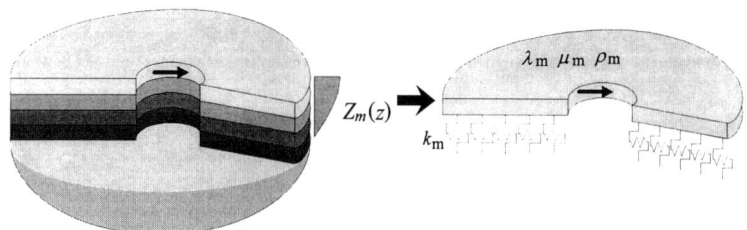

(a)   soil deposit surrounding an embedded body   (b) soil plane overlying Winkler springs

Figure. 3 Simple ground model for a particular vibration mode with respect to depth

pattern of displacement, respectively to those of the model. They are expressed in terms of $Z_m(z)$ as:

$$\lambda_m = \int_0^L \lambda(z) \cdot Z_m(z)^2 \, dz, \quad \mu_m = \int_0^L \mu(z) \cdot Z_m(z)^2 \, dz, \quad \rho_m = \int_0^L \rho(z) \cdot Z_m(z)^2 \, dz \quad \text{and}$$

$$k_m = \int_0^L \mu(z) \left(\frac{\partial Z_m(z)}{\partial z} \right)^2 dz = \rho_m \cdot \omega_m^2 \qquad (2a)\text{-}(2d)$$

The soil stiffness S_x for this particular motion $Z_m(z)$ of a cylindrical cavity in the surface layer is then expressed as the stiffness for the lateral motion of a rigid massless disk in the model's plane of a unit thickness: which is found completely identical to that given in Novak et al (1978) regardless of the presence of Winkler-type springs, i.e:

$$S_{x,m} = \pi \mu_m a_m^2 \frac{4K_1(b_m)K_1(a_m) + a_m K_1(b_m)K_0(a_m) + b_m K_0(b_m)K_1(a_m)}{b_m K_0(b_m)K_1(a_m) + a_m K_1(b_m)K_0(a_m) + b_m a_m K_0(b_m)K_0(a_m)} \qquad (3)$$

The only difference from Novak's solution, owing to the presence of Winkler-type springs, k_m, appears in the dimensionless frequencies a_m and b_m as:

$$a_m = \frac{\omega_m r_0}{v_T} \eta_m, \quad b_m = \frac{\omega_m r_0}{v_L} \eta_m, \quad \omega_m = \sqrt{\frac{k_m}{\rho_m}} \quad \text{and} \quad \eta_m = \sqrt{(1+iD) - \left(\frac{\omega}{\omega_m} \right)^2}$$

$$(4a)\text{-}(4d)$$

in which, v_T and v_L are transverse and longitudinal wave velocities in the plane, respectively. The expression for a Poisson's ratio v_m equal to 0.5 is obtained by taking a limit as $v_m \rightarrow 0.5$ in equation (3), i.e.:

$$S_{x,m} = 2S_m^* + m_s (\omega_m \eta_m)^2 \qquad (5)$$

where, $m_s (= \rho_m \pi r_0^2)$ is the soil mass of the same volume as the rigid mass-less disk in the soil plane, and

$$S_m^* = 2\pi \mu_m \frac{a_m K_1(a_m)}{K_0(a_m)} \qquad (6)$$

Nogami and Konagai (1988, 1994) have found that the Novak's solution $S_{x,m}$ (equation (3)) for the Poisson's ratio other than 0.5 can be approximately expressed in the same form as equation (5) but with a small modification, i.e.:

$$S_{x,m} = \xi(v_m) \cdot S_m^* + \zeta(v_m) \cdot m_s (\omega_m \eta_m)^2 \qquad (7)$$

where, $\xi(v_m)$ and $\zeta(v_m)$ are functions dependent only on a Poisson's ratio. The values $\xi(v_m)$ and $\zeta(v_m)$ are given in Table I. As has been already mentioned, this approach is based on the assumption of vanishing vertical displacement, and this assumption leads to assuming plane strain condition over the entire extent of the soil plane. However, Konagai et al. (1992) have shown that assuming plane stress condition allows $S_{x,m}$ to approximate more closely the rigorous solution of soil stiffness, because the stress free condition on the ground surface affects greatly the motion of the entire surface layer. To demonstrate the validity of assuming the plane-stress condition, Konagai (1992) has conducted a model experiment to visualize contour lines of lateral displacement caused by the transient motion of an embedded cylinder. A soft soil model, made up of transparent poly-acrylamide gel, was prepared in an acrylic box (57 cm ×57 cm ×13.5 cm). An acrylic cylinder with a radius of 50 mm was embedded in it as shown in **Figure 4**. This cylinder has a small cone of brass on its base so that the cylinder can rock on the bottom of the cylinder. Flexible black stripes of 0.5 mm thickness were printed at an interval of 1.0 mm on the model's surface, and an impulse was given to the top of the cylinder in the horizontal and normal direction to the stripes. Stripes before and after excitement were photographed and superimposed on one frame of film so that the moiré fringe would appear. **Figure 5** shows the observed moiré patterns. A shear wave propagates in the orthogonal direction to the excitement, and another different clear wave front is observed in the longitudinal direction. Since Poisson's ratio of the gel, cured in water, is about 0.5, actual longitudinal wave velocity must be extremely large in comparison with the shear wave velocity. The visualized wave front, however, propagates at about double the speed of the shear wave velocity, which is identical to the longitudinal wave velocity through a plane-stress medium. Later on, Mikami and Konagai (1994) have succeeded in taking similar pictures even in the interior of the soil model, which clearly show that even the interior of the soil model is strongly affected by the presence of the stress-free ground surface.

Poisson's ratio v_m in equation (7), thus, must be replaced with v_m^* for a plane-stress medium, which is expressed as:

Table I   Values of ξ and ζ

Poisson's ratio, v	ξ	ζ
0.50	2.000	1.0000
0.47	1.831	0.5336
0.45	1.741	0.3740
0.43	1.667	0.2628
0.40	1.580	0.1428
0.35	1.476	0.0352
0.25	1.351	0
0.20	1.311	0
0.10	1.252	0
0.00	1.213	0

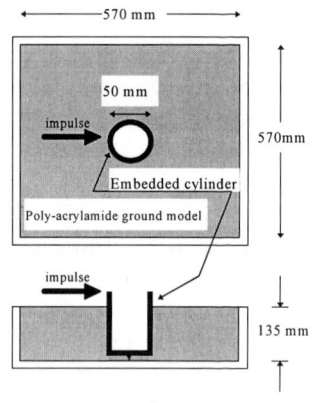

Figure 4   Embedded structure model

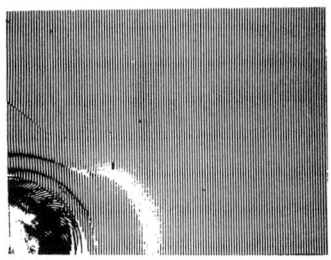

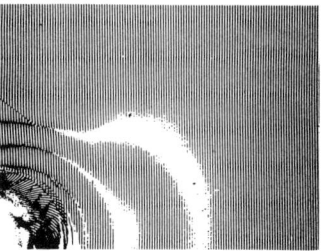

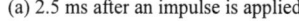

(a) 2.5 ms after an impulse is applied   (a) 5.0 ms after an impulse is applied

Figure 5   Observed Moiré fringes

$$v_m^* = \frac{\lambda_m^*}{2(\lambda_m^* + \mu_m)} \quad \text{where,} \quad \lambda_m^* = \frac{2\lambda_m \mu_m}{\lambda_m + \mu_m} \qquad \text{(8a), (8b)}$$

It is noted that the modified Poisson's ratio v_m^* never exceeds 1/3, and thus, $\zeta(v_m^*)$ in equation (7) is always equal to zero. Equation (7) is then rewritten as;

$$S_{x,m} = \xi(v^*) \cdot S_m^* \qquad (9)$$

Inverting the stiffness $S_{x,m}$ in equation (9) yields the flexibility function $H_{x,m}$ ($= 1/S_{x,m}$), which is found to be closely approximated by the following form as:

$$H_{x,m}(s) = A_{e,m}\frac{1}{s + \alpha_{e,m}} + A_{c,m}\frac{s + \alpha_{c,m}}{(s + \alpha_{c,m})^2 + \omega_m^2} \qquad (10)$$

where, $s = i\omega$, $A_{e,m} = A_{c,m} \cdot \left(\frac{2L}{r_0} - 1\right)$, $A_{c,m} = \left(0.44 - 0.04 \cdot \log_2 \frac{L}{r_0}\right) / \xi(v_m^*) / 2\pi\mu_m$,

$\alpha_{e,m} = \left(1.2 + 0.4 \cdot \log_2 \frac{L}{r_0}\right) \cdot \omega_m$ and $\alpha_{c,m} = \left(0.21 - 0.01 \cdot \log_2 \frac{L}{r_0}\right) \cdot \omega_m$

(11a)-(11e)

Figures 6a and 6b show flexibility functions of a cylindrical hollow undergoing the vibration mode $Z_m(z)$ for different surface thickness-radius ratios (L/r_0) 2 and 8, respectively. Within the range of thickness-radius ratio $1 < L/r_0 < 10$, the approximate expression agrees well with the rigorous solution. Inverse Fourier transformation of $H_{x,m}$ yields the impulse response function $h_{x,m}(t)$ as:

$$h_{x,m}(t) = A_{e,m}h_{e,m}(t) + A_{c,m}h_{c,m}(t) \qquad (12)$$

where, $h_{e,m} = e^{-\alpha_{e,m}t}$, $h_{c,m} = e^{-\alpha_{c,m}t}\cos\omega_m t$ (13a), (13b)

It is now clear from equation (12) that the impulse response function $h_x(t)$ for any vibration mode is approximated by a linear combination of exponential and exponentially decaying cosine functions.

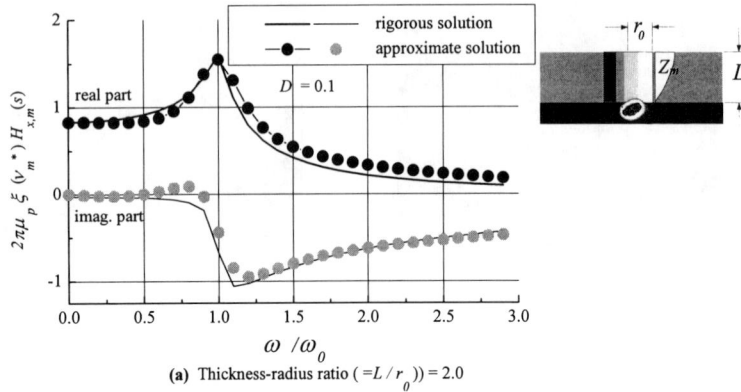

(a) Thickness-radius ratio ($=L/r_0$)) = 2.0

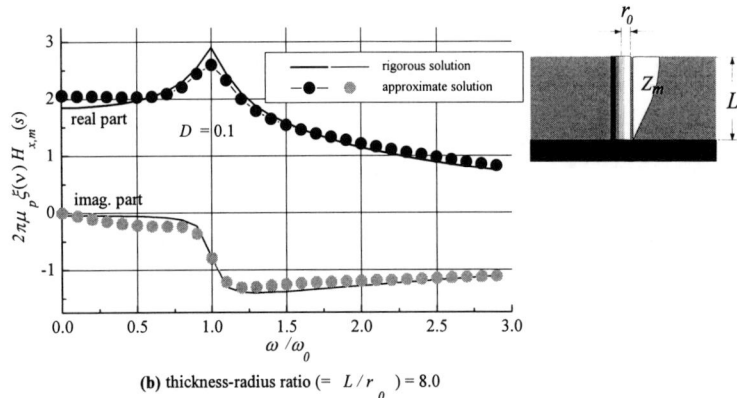

(b) thickness-radius ratio (= L/r_0) = 8.0

Figure. 6 Flexibility functions $H_{x,m}(s)$ for different thickness-radius ratios

Soil-Structure Interaction at the Base

Meek and Wolf (1992a-1993b) have developed a unified approach for soil-structure interaction analysis by using truncated semi-infinite cone models representing an unbounded soil medium. By superimposing contributions of all the mirror images of the cone, their approach covers a wide variety of soil-structure systems including surface foundations, embedded bodies, flexible piles and so on. However, the very basic concept is found in a simple rigid circular foundation on the surface of a homogeneous soil half-space: which can be viewed exactly as an appropriate example explaining the soil-structure interaction at the base of an embedded foundation. According to their approach, the soil is idealized for each degree of freedom as a truncated semi-infinite elastic cone with its own apex height z_0 (**Figure 7**). The apex ratio z_0 / r_0, or the opening angle of the cone, is determined just by equating the static stiffness coefficient of the disk on the semi-infinite soil half-space to that of the corresponding cone: whereas the wave propagating through the cone with the velocity v dominates the stiffness within the considerably high frequency range. For a translational cone, the unit-impulse response function $h_x(t)$ is obtained as:

$$h_x(t) = \begin{cases} \dfrac{1}{K_{x,static}} \dfrac{v_T}{z_0} e^{-\frac{v_T}{z_0}t} & t > 0 \\ 0 & t < 0 \end{cases} \tag{14}$$

with $K_{x,static} = \rho v_T^2 \cdot \pi r_0^2 / z_0$. When rocking motion of the disk (moment of inertia: $I_0 = (\pi / 4) r_0^4$) is concerned, a rotational cone is to be discussed. The unit-impulse response function $h_\theta(t)$ for a rotational cone is obtained as:

$$h_\theta(t) = \begin{cases} \dfrac{1}{K_{\theta,static}} \dfrac{v_L^*}{z_0} e^{-\frac{3v_L^*}{2z_0}t} \left(3\cos\dfrac{\sqrt{3}}{2}\dfrac{v_L^*}{z_0}t - \sqrt{3}\sin\dfrac{\sqrt{3}}{2}\dfrac{v_L^*}{z_0}t \right) & t > 0 \\ 0 & t < 0 \end{cases} \tag{15}$$

where, $K_{\theta,static} = 3\rho v_L^{*2} I_0 / z_0$

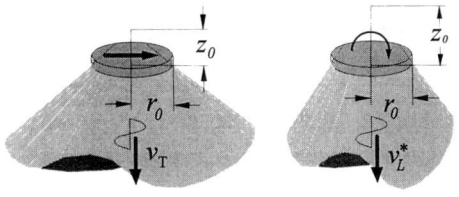

translational cone   rotational cone

Figure 7   Cones for various degrees of freedom (Meek and Wolf (1992a-1993b))

<u>Analog Circuit</u>

All above-mentioned models may be such an oversimplification of reality that they can not cover all cases. They, however, allow the impulse response function $h(t)$ for any of lateral, vertical or rotational vibration modes to be approximated by linear-combinations of basic response functions $h_m(t)$ as:

$$h(t) = \sum_{m=1}^{n} A_m h_m(t) \tag{16}$$

where, A_m is an unknown constant and

$$h_m(t) = \begin{cases} e^{-\alpha_m t}\cos(\omega_m t - \phi_m) & t \geq 0 \\ 0 & t < 0 \end{cases} \tag{17}$$

The basic response functions expressed by equation (17) are simple exponential functions and/or exponentially decaying oscillations. Setting ϕ_m in equation (17) at $\pi/2$ yields an exponentially decaying sine function: which can be completely conformed to the simple-damped oscillation of any degree of freedom of a structure. Thus, these functions can also be used when a part of a structure or an attached device alone, like a tuned-mass damper on or in a building for example, is tested. Fourier transform of $h_m(t)$ in equation (17) is:

$$\mathfrak{F}(h_m(t)) = H_m(s) = \frac{s\cdot\cos\phi_m + (\alpha_m\cos\phi_m - \omega_m\sin\phi_m)}{s^2 + 2\alpha_m s + \alpha_m^2 + \omega_m^2} \tag{18}$$

where, $s = i\omega$, and $\mathfrak{F}$ is the abbreviation of Fourier transformation. Needless to say, stiffness expression in the frequency domain is just an inverse of equation (18).

It is noted here that both the flexibility (equation (18)) and stiffness expressions in the frequency domain have the following common form as:

$$H(s) \text{ (or its inverse } 1/H(s)) = \frac{a_0 + a_1 s + a_2 s^2}{b_0 + b_1 s + b_2 s^2} = \frac{e_o}{e_i} \tag{19}$$

where, e_i and e_o are input and output signals, respectively. Electric signals can be controlled by using analog circuits. The first-level units in analog circuits are operational amplifiers and passive elements (resisters and capacitors, respectively) (See Holman, 1978, for example). These units form a linear amplifier (**Figure 8a**), an adder (**Figure 8b**), and an integrator (**Figure 8c**), which are respectively to add several different signals together, to multiply an input signal by a scale factor a, and to integrate an input electric signal. An adder, a linear amplifier and an integrator are

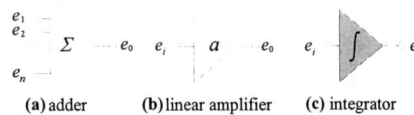

(a) adder   (b) linear amplifier   (c) integrator

Figure 8   Key circuits

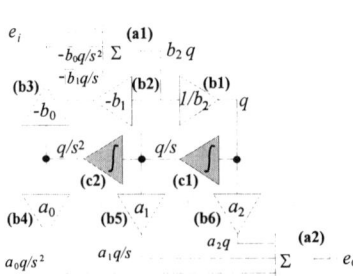

Figure 9 Analog circuit to generate basic response functions

Figure 10 Basic response function generator

the key circuits used in designing analog electric circuits.

Introducing an unknown quantity q, the above equation (19) can be separated into the following two equations as:

$$e_o = a_0 \frac{q}{s^2} + a_1 \frac{q}{s} + a_2 q \text{ and } e_i = b_0 \frac{q}{s^2} + b_1 \frac{q}{s} + b_2 q \qquad \text{(20a), (20b)}$$

With the expression in equations (20a) and (20b), the circuit that is capable of generating e_o to an arbitrary input signal e_i is designed as shown in **Figure 9**. The input signal e_i and two additional signals, later defined to be $-b_0 \cdot q / s^2$ and $-b_1 \cdot q / s$, are added together first by the adder (**a1**) and then multiplied by $1/b_2$ by the linear amplifier (**b1**). The output signal in the above process is q according to equation (20b). Noting that integrating a signal is equivalent, in the frequency domain, to divide its Fourier spectrum by s, integrators (**c1**) and (**c2**) produce signals q/s and q/s^2, respectively. After these two signals go through linear amplifiers (**b2**) and (**b3**) with scale factors $-b_1$ and $-b_0$ respectively, they become $-b_1 \cdot q / s$ and $-b_0 \cdot q / s^2$, and returned to the adder (**a1**): whereas linear amplifiers (**b4**), (**b5**) and (**b6**) produce $a_0 q / s^2$, $a_1 q / s$ and $a_2 q$ respectively, and they are added together by the adder (**a2**). It is now clear from equation (20a) that the output of the adder (**a2**) is identical to the signal e_o.

Figure 10 shows a model for a test try of **Figure 9**-equivalent circuit. Since the circuit model is designed just to simulate flexibility functions (equation (17)) only, the linear amplifier (**b6**) with the scale factor a_2 is not built in. Five pairs of knobs are for tuning the five scale factors in **Figure 9**. In **Figure 11** examples are shown of transient response of the circuit to an impulse (rectangular pulse of 5V, duration time = 10 ms). Only tuning the parameters to prescribed values allows any of the basic response functions to be generated.

Since equation (16) implies that an impulse response function for any of lateral, vertical or rotational vibration modes can be approximated by a linear-combination of basic response functions generated by the present circuit, a necessary number of the circuits should be wired up so that an input signal p is applied to each one of these sub-circuits and outputs from them are added together by an adder (**Figure 12**).

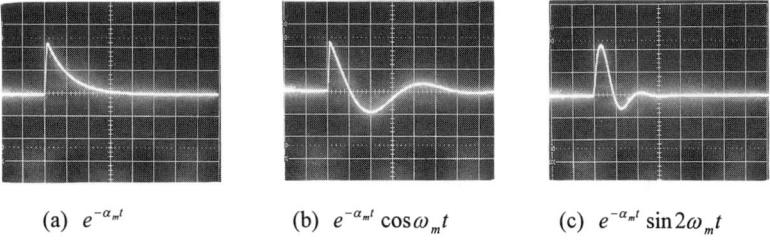

(a) $e^{-\alpha_m t}$ (b) $e^{-\alpha_m t} \cos \omega_m t$ (c) $e^{-\alpha_m t} \sin 2\omega_m t$

Figure 11 Basic response functions generated by the present analog circuit (0.1 s/div.)

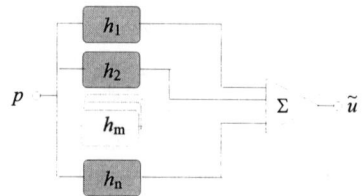

Figure 12   Assembly of h_m (m=1, 2,....n)

Stability of the Present System for Controlling Shaking Tables

The assemblies of sub-circuits (**Figure 12**), that are designed to generate unit-impulse response functions for lateral, vertical and rocking motions, are used to produce the outputs corresponding to the soil-structure interaction motions, $\tilde{u}_x$, $\tilde{u}_z$ and $\tilde{u}_\theta$ in **Figure 1**. This system functions in the intended manner, to be sure, however when a less-damped structure model is tested, unexpected noise amplification causes serious problem in operating a shaking table. **Figure 13** shows the basic signal flow of the present system in which u and $\tilde{u}$ are the free-field ground motion and the soil-structure interaction motion, respectively. A structure model, responding to the input ground motion $u + \tilde{u}$, is expected to produce interaction forces $H_0 \cdot (u + \tilde{u})$. In actuality however, the motion of a shaking table normally contains some steady noise n, which is added together with $u + \tilde{u}$. Therefore, after the actual signal $H_0 \cdot (u + \tilde{u} + n)$ goes through the present analog circuit with a transfer function H_1, it becomes $H_1 \cdot H_0 \cdot (u + \tilde{u} + n)$ that is exactly identical to the interaction motion $\tilde{u}$. Equating $H_1 \cdot H_0 \cdot (u + \tilde{u} + n)$ to $\tilde{u}$ results in:

$$\tilde{u} = \frac{H(u+n)}{1-H} \tag{21}$$

where,

$$H = H_1 \cdot H_0 \tag{22}$$

Equation (21) shows that the interaction motion $\tilde{u}$ can be seriously affected by the presence of the noise n when $|H| \gg |1 - H|$.

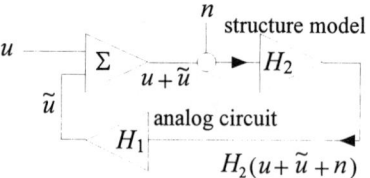

Figure 13   Signal flow for controlling
shaking table

An upright flexible cantilever (a steel strip: 2,000 mm × 300 mm × 8 mm) with an extremely small damping constant of only 0.2% (**Figure 14**) is tentatively put on a shaking table (Konagai and Katsukawa, 1997). The model is assumed to be supported by a circular surface foundation ($v_T / r_0 = 10 \, s^{-1}$, $K_{x,static} = 16.4$ kgf/cm, $K_{\theta,static} = 1.62 \times 10^5$ kgf·cm) whose unit-impulse response functions for sway and rocking motions are obtained by equations (14) and (15). An impulse shown in **Figure 15** is given to the shaking table as an input motion u. The dotted line in **Figure 16** shows the response of the model's top end to the input motion u without the interaction motion $\tilde{u}$ being added: whereas the thin line shows the response affected by the interaction motion $\tilde{u}$. Incorporating the effect of the interaction motion leads to the increase of damping up to 2.2% and to the slight decrease of natural frequency as well. However, what we should notice among all the features observed in this figure is the serious noise that has been intentionally left unfiltered in the time interval from 7 to 8 seconds. The predominant frequency of the noise is about 11 Hz, and is about identical to the fourth natural circular frequency of the model. This noise having the nearly constant and a bit higher frequency will be reduced to a great extent if it is integrated twice to become the time history of displacement. The noise, however, can be larger and more serious depending on soil-foundation systems to be studied. Thus, the transfer function $H (= H_0 \cdot H_1)$ must be

Figure 14 Upright beam on a shaking table

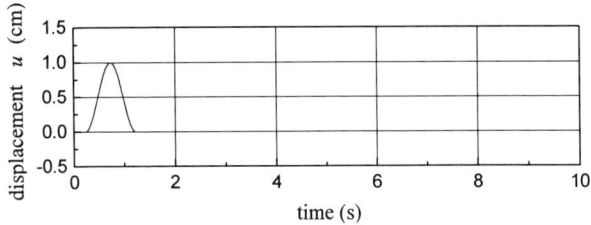

time (s)

Figure 15   time history of applied displacement

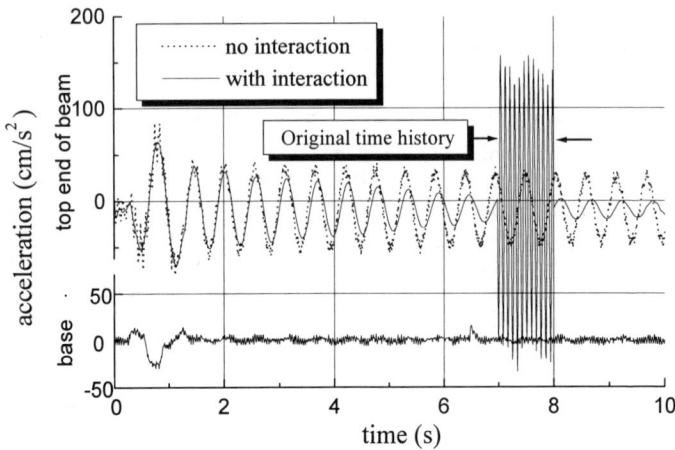

Fig. 16   Acceleration responses at top and bottom ends of beam

foundation systems to be studied. Thus, the transfer function $H \, (= H_0 \cdot H_1)$ must be examined beforehand to avoid bringing the value of H close to 1.

Conclusions

A new method for a model experiment on a shaking table has been presented. The present method allows soil-structure or base-structure interaction to be simulated. The conclusions of this study are summarized as follows:

(1) Unit-impulse response functions at the side or at the base of an embedded body for lateral, vertical and rotational vibration modes are closely approximated by linear combinations of basic response functions, namely, exponential and/or exponentially decaying oscillations.

(2) Any of the basic response functions is produced by one analog circuit with five parameters being adjusted to the prescribed values.

(3) The above-mentioned circuit enables us to simulate the soil-structure interaction in model tests by using a structure model alone without a physical ground model. Since analog circuits lose hardly any time in producing output signals, they can be used in a shaking table test which is operated under relatively fast loading in earthquake simulation.

(4) When a less-damped structure model is tested on a shaking table, unexpected noise amplification can cause serious problem in operating the shaking table. When the soil-structure interaction motion $\tilde{u}$ is given in the present system as a

the amplification factor is identical to $H / (1 - H)$. This transfer function must be examined beforehand to avoid bringing the value of H close to 1.

Acknowledgment

Partial financial support for this study has been provided by the Foundation of Promoting Industrial Science, IIS, University of Tokyo. Grateful acknowledgment is made to Dr. Takeyasu Suzuki, Technical Research and Development Institute, Kumagai Gumi, Co., Ltd., for his kind advice and help throughout the course of our study. The authors are indebted to Mr. Tota Katsukawa, Technical Research and Development Institute, Kumagai Gumi, Co., Ltd. and Mr. Toshihiko Katagiri, IIS, Univ. of Tokyo, for their help throughout the experiments.

References

Holman, J. P. (1989). *"Experimental Methods for Engineers,"* McGraw-Hill.

Konagai, K. and Maehara, M. (1992). "Study on hypotheses for simple numerical evaluation of soil-embedded structure interaction," *Bull., Earthquake Resistant Structure Research Center*, **25**, 39-60.

Konagai, K. and Nogami, T. (1994). "Subgrade model for transient response analysis of multiple embedded bodies," *Earthquake eng. struct. Dyn.* **23**, 1097-1114.

Konagai, K. and Katsukawa, T. (1977). "Real time control of a shaking table for simulation of soil-flexible structure interaction," *Bull., Earthquake Resistant Structure Research Center*, IIS, Univ. of Tokyo, **30**.

Nogami, T. and Konagai, K. (1986). "Time-domain axial response of dynamically loaded single pile," *J. eng. mech. ASCE*, **112(11)**, 1241-1252.

Nogami, T. and Konagai, K. (1988). "Time-domain flexural response of dynamically loaded single piles," *J. eng. mech. ASCE*, **114(9)**, 1512-1525.

Novak, M. and Nogami, T. (1977). "Soil-pile interaction in horizontal vibration," *Earthquake eng. struct. Dyn.* **5**, 263-281.

Novak, M., Nogami, T. and Abouli-Ella (1978). "Dynamic soil reactions for plane strain case," *J. eng. mech. ASCE* **104**, 953-959.

Meek, J. W. and Wolf, J. P. (1992a). "Cone models for homogeneous soil," *J. geotechnical eng., ASCE*, **118(5)**, 667-685.

Meek, J. W. and Wolf, J. P. (1992b). "'Cone models for soil layer on rigid rock," *J. geotechnical eng., ASCE*, **118(5)**, 686-703.

Meek, J. W. and Wolf, J. P. (1992c). 'Cone models for embedded foundation', *J. geotechnical eng., ASCE*, **120(1)**, 60-80, (1992c).

Meek, J. W. and Wolf, J. P. (1993a). "Cone models for nearly incompressible soil," *Earthquake eng. struct. Dyn.,* **22**, 649-663.

Meek, J. W. and Wolf, J. P. (1993b). "Why cone models can represent the elastic half space," *Earthquake eng. struct. Dyn.,* **22**, 759-771.

Mikami, A. K. Konagai, J. Sayama and T. Tamura: Simple Approach for Evaluation of Dynamic Cross-Interaction between Closely-Spaced Embedded Structures, *9th Japan Earthquake Engineering Symposium*, Vol. 3, E277-E282, 1994.

Japan Earthquake Engineering Symposium, Vol. 3, E277-E282, 1994.

Tajimi, H. (1969). "Dynamic analysis of a structure embedded in an elastic stratum," *Proc. 4th World Conf.*, *Earthquake Engineering*, Santiago, Chile, **III(A-6)**, 53-69.

Wolf, J. P. (1988). "Dynamic soil-structure interaction analysis in time domain," Englewood Cliffs, NJ: Prentice-Hall.

Development of a Large-scale Strong Ground Motion Test Facility
for Studying Structural Response and Soil-Structure Interaction

Pete Mote, Paul Gefken, Don Curran, Dave McCallen,
Ignacio Arango, and Orhan Gurbuz
[1]Nevada Testing Institute (NeTI)

Abstract

This paper presents an update on an ongoing effort by the Nevada Testing Institute (NeTI) to design a soil cantilever (soil island) test bed suitable for studying structural response and soil-structure interactions arising from near-source earthquakes. The earthquake-like ground motion will be generated using SRI International's RESCUE (Repeatable Earth Shaking by Controlled Underground Expansion) technique by loading a soil cantilever test bed on one or more sides with a series of controlled pressure pulses. The pulses are applied by expanding sources driven by propellant gas generators. A smaller (1/5 of the size) test bed is planned for construction this year at the Nevada Test Site (NTS) in an area that is also suitable for eventual full-scale test bed construction.

The main objectives of the smaller test is to evaluate RESCUE hardware performance and to validate computer models used to predict test-bed response. Ultimately, this smaller test bed will be used repeatedly to test a variety of structures, including piles and structures at full-scale. The smaller soil cantilever test bed will have dimensions on the order of 1.2-meters wide by 9-meters long by 4.6-meters deep. Based on the results of tests on this test bed, a facility large enough to test large structures will be designed and constructed at NTS. The full-scale soil cantilever test bed will have dimensions on the order of 46-meters square by 23-meters deep. This paper describes the desired response spectra for the testing facility, NTS site geotechnical characteristics, soil cantilever and RESCUE source module requirement calculations for the test bed, design, construction, and testing schedule of the smaller RESCUE source module.

Background

Recent strong motion earthquakes in California, Japan, and elsewhere around the world have dramatically demonstrated the urgent need for better large-scale earthquake simulation and seismic testing of large-scale structures. This need

---

Nevada Testing Institute, Inc., 755 E. Flamingo Road, P.O. Box 19360, Las Vegas, NV 89132

is particularly evident when there is relative motion between the structure and surrounding soil, such as when structures are located in regions where soil liquefaction occurs. Current seismic simulation testing methods do not provide solutions to all earthquake-related problems because they typically do not include the soil-structure interaction, do not use large-scale, complete structural systems, and do not apply the loads to the structure through a soil foundation. The NeTI design team, which consists of engineers and scientists from the Bechtel Corporation, Defense Special Weapons Agency, Lawrence Livermore National Laboratory, Los Alamos National Laboratory, Sandia National Laboratory, and SRI International, is addressing this problem by developing a specialized testing facility that allows for better large-scale earthquake simulation and testing of large-scale structures with the inclusion of soil-structure interaction. The goal of the NeTI design team is to develop a testing facility that produces earthquake-like ground motions with significant durations between 10 and 20 seconds, and that meet or exceed the response spectra shown in Figure 1 to allow for studying structural damage and soil-structure interaction behavior. These ground motions can be tailored to meet specific structural engineering needs.

The Nevada Testing Institute (NeTI)

Bechtel Nevada (BN); a partnership between Bechtel Nevada Corporation, Johnson Controls Nevada Corporation, and Lockheed Martin Nevada Technologies Corporation, began managing and operating the Nevada Test Site (NTS), the Department of Energy's weapons testing facility in Southern Nevada on January 1, 1996. A significant aspect of BN's management and operations strategy for NTS, is to make available the broad scientific and testing capability of the NTS to benefit the public and science communities. Because of NTS's deep roots in science, testing, and diagnostics, NTS is an ideal location to develop an international testing laboratory where unique experiments benefiting all aspects of civil and structural engineering can be conducted.

In keeping with this strategy, BN, the National Advanced Drilling and Excavation Technologies Institute (NADET) at the Massachusetts Institute of Technology, and SRI International (SRI), established the non-profit institute, NeTI, for applied research, technology development, testing, and demonstration of advanced underground, civil, structural, and seismic engineering technologies. NeTI has its headquarters in Las Vegas and conducts its field operations at NTS, some 65 miles north of Las Vegas.

NeTI is applying NTS capabilities, testing culture, infrastructure, physical character, and remoteness to advance the state of practice of civil, structural and seismic engineering, capitalizing on the unique explosive technologies capabilities and experience that exist within the NeTI team and the NTS complex. NeTI's full scale testing facilities at NTS will support advancement of structural designs to withstand loads from earthquakes forces, including base isolation and ground modification techniques. They will integrate with existing seismic engineering testing facilities as the next step in the structural engineering testing process.

Existing Seismic Simulation Testing Methods

Four major methods are currently available for determining the response of prototype-size structures subjected to earthquake loading: quasi-static testing, pseudo-dynamic testing, shaking table testing, and earthquake monitoring. These methods all have certain advantages and disadvantages.

Quasi-static testing of structural components and subassemblies is economical and simplified. Using this method, large-scale structural components are slowly subjected to assumed earthquake loads (quasi-static cycling) by placing concentrated loads at predetermined locations. This method has advantages for studying local detailed phenomena, such as detailing aspects, connection regions, and the load-deformation behavior of individual elements. The main disadvantages of this method are the limitations in accurate modeling of the boundary and loading conditions, the lack of inclusion of the response of a complete structural system, and the lack of dynamic effects.

The pseudo-dynamic testing technique overcomes some of the disadvantages of the quasi-static test method. In applying this technique, a large scale model structure is constructed in a laboratory equipped with strong reaction walls and floor. To model major earthquake loads that come from inertial reaction loads, dynamic loading actuators are placed at concentrated masses. The energy dissipation of the load-carrying members occurs physically in the laboratory, and the loads that are applied represent the mass dynamic response effects that are modeled in a hybrid digital computer simulation. The disadvantages of the technique are that complex mechanical hardware is needed (especially for multiple degree of freedom structures and excitations), the structure undergoes non-realistic specimen damping effects, and potential numerical integration instability may occur in the algorithms that compute the applied inertial loads.

Shaking table testing is the most realistic simulation of actual earthquake base motion events. However, the capital equipment costs of a shaking table are an exponential function of specimen weight and dimension, so specimens must be reduced in scale for economy. This makes it very difficult to scale some types of structural behavior, such as shear and bond strength in reinforced concrete elements. Another economic factor is that a rigid controllable structural base or table must be carefully designed and manufactured. A realistic rule of thumb is that the table weighs as much as the testable specimen. A particularly difficult problem is the evaluation of soil-structure interactions from seismic excitations because of the large volume of soil necessary on the shaking table.

The earthquake monitoring method consists of studying structural response resulting from an actual earthquake. This has been one of the most effective methods for understanding the response of complete structural systems including the surrounding soil. However, the time and location of any given earthquake are unpredictable. The cost of assuring that recording and measuring devices will function at the time of a seismic event is high and often difficult to justify when one considers that decades may pass before the equipment is called into use.

The NeTI Strong Ground Motion Generator

The NeTI Strong Ground Motion Generator will be located NTS. NTS site and national laboratory personnel will provide whatever support deemed necessary by the user for running the test facility, construction of large scale test structures, and measurement of ground motion and structural response. The soil cantilever test bed, which can be viewed as a large soil-shaking table, will be approximately 46-meters square and will be surrounded by a 23-meter deep trench as shown in Figure 2. Test bed ground motion will be generated using the RESCUE technique, which consists of specially designed buried and reusable propellant sources that expand and contract to produce multiple pressure pulses on the test bed side.

The RESCUE technique has been developed and evaluated through numerous numerical simulations and large and small-scale tests (1-7). As will be shown later, the RESCUE technique, combined with a tailored soil cantilever test bed, can provide realistic seismic response spectra for testing structures. The NeTI Strong Ground Motion Testing Facility applys its dynamic loads to the structure through the surrounding soil, thus, allowing for the study of the response of large-scale structures with the inclusion of soil-structure interaction.

The RESCUE technique produces ground motion by simultaneously expanding a planar array of buried vertical sources. These expanding sources move the soil, which excites the dynamic response of the structure. Because the sources do minimal damage to the surrounding soil (applied pressure is less than 1 MPa), sequential pulses of ground motion can be applied, and follow-on tests can be conducted at the same location and on the same structure. The RESCUE source design shown in Figure 3 consists of a neoprene-rubber bladder around a rectangular mandrel. The steel canisters contain propellant that produces high pressure gas when ignited. This gas is vented into the source module in a controlled manner, causing expansion of the rubber bladder against the soil cantilever. When the bladder expands, it moves the soil. The source is surrounded by a steel frame that prevents failing of the soil at the bladder-soil interface. Individual sources are lined up side-by-side to apply a planar load to the test bed side. Each source can contain between 10 to 20 canisters to produce up to 20 pulses of ground motion.

The RESCUE technique allows the characteristics of the generated pressure pulse to be easily controlled. The peak pressure depends on the initial pressure within the steel canister, which is a function of the propellant quantity. The rise time of the pulse depends on how quickly the gas from the canisters is vented into the rubber bladder, and the pulse duration depends on the timing of the release of pressure from the rubber bladder to the atmosphere.

To illustrate the anticipated capabilities of the NeTI Strong Ground Motion Testing Facility, we performed numerical plane strain finite-element calculations using DYNA3D. Figure 4 shows a cross section of the numerical model and the multiple pressure pulses applied to the side of the test bed. The calculations were performed using the Ramberg-Osgood elasto-plastic material model (8). The following test bed soil properties, which are considered to be representative of the NTS alluvium, were adopted for use in the DYNA3D model:

- Density = 2000 kg/m3
- Shear Modulus = 335 MPa
- Shear Wave Velocity = 400 to 500 m/s
- Mohr-Coulomb Friction Angle = 40 degrees
- Poisson Ratio = 0.35

The calculation indicates that the soil cantilever model has a natural frequency of 2.4 Hz. Figure 5 shows the calculated response spectra. Between the approximate corner frequencies of 8 Hz to 2 Hz, the NeTI Strong Ground Motion Testing Facility will meet or exceed the design response spectra shown in Figure 1.

NTS - Geotechnical Characterization

Due to the considerable extent of the NTS, subsurface soil conditions vary from one locality to another. There are rock exposures, and colluvial, alluvial, and lake deposits. However, the most of the locations of interest for siting the Test Facility consist of deep alluvial sand and gravel deposits.

Typically, the sands and gravels are loose to medium dense at shallow depths, with density increasing at lower elevations. Fines content is low, less than about 20 percent, and the maximum gravel particle size is generally less than 15 to 20 cm. The ground-water table elevation varies but is generally lower than 30 meters. Only limited geotechnical test data are available from the alluvium. This is due to the granular nature of the deposits.

Recently, a shallow geophysical survey was completed at the location of one of the candidate sites for the facility (9). Three geophysical survey lines were tested using P-wave and S-wave generators. The test results showed that for the upper 10 meters, the average compressional and shear wave velocities were near 600 m/s and 400 m/s, respectively. Based on geologic and deep-trench observations, other alluvial sites may have lower wave propagation characteristics.

For the purpose of the Test Facility, it is desirable that the cantilever test bed material be sufficiently strong to support the stresses imposed by the expanding bladder, yet soft enough not to filter out the low-frequency ground motions. It is also desirable for it to have self-healing properties, i.e., the capability of closing fissures/cracks that may result from the ground shaking. It appears that the alluvial deposits at the NTS exhibit these desirable properties.

Small Demonstration Test

A series of small demonstration tests are planned at NTS in the summer of 1997. The objectives for these tests is to checkout the RESCUE source hardware, evaluate the test bed soil behavior, and validate the DYNA3D finite element model used to calculate system performance.

A scaled version of a RESCUE source, referred to as SRISM (SRI source module), will be used to load the soil cantilever test bed. SRI has fabricated a

3-meter long source that has the capability of producing two pulses. It is estimated that the SRISM will produce peak test bed displacements of about 2.5 to 5.0 cm. Plans for the source are shown in Figure 6. Figure 6a shows the source module that supports the expandable neoprene-rubber bladder and contains the propellant canisters. The screen on the outside of the source module is 1.3 cm #18 standard expandable metal. It acts as a mandrel that supports the rubber bladder when the source is not pressurized and allows the pressurized gas from the canisters to pass through and expand the rubber bladder. The two fixtures that extend from the top of the source hold 30.5-cm-diameter helium controlled exhaust valves that allow for venting of the bladder and are used to control the pulse duration. The 2.54 cm thick neoprene-rubber bladder for the SRISM will be fabricated and installed by the Goodyear Rubber Co. of Rancho Cucamonga, Calif.

The source contains two steel canisters that have an outside diameter of 12.7 cm and a wall thickness of 1.9 cm. The canisters shown in Figure 6b were fabricated using 4130 steel, which has a yield strength of 703 MPa and a maximum elongation of 21%. These canisters can hold up to about 4 kg of propellant and produce SRISM internal pressures of up to 1 MPa. Each canister has seven vents evenly spaced along its length. The vents are designed to control the rise time of the pressurization (by the diameter of the through holes) and to deflect the flow of the pressurized gas away from the rubber bladder and the adjacent canister.

The source module is surrounded by a steel support frame. The support frame confines the rubber bladder at the edges. The top of frame is designed to have adequate strength to contain the pressure and to be disassembled for access to the canisters and backfilling of soil around the source. When an array of sources is used to load the large soil test bed, the support frame needs only to confine the end sources and the top of each source. Within the array, adjacent sources provide side confinement for each other.

Figure 7 shows the SRISM hardware schematic diagram. The timing and firing unit operates the timing for the propellant ignition using the electronic detonation units (EDU's) and the venting of the gas pressure within the SRISM. The SRISM is purged with nitrogen before the test so that no after burning of the propellant outside the canisters occurs. The combination of propellant quantity, venting of high pressure gas into the source, and venting of the low pressure gas from the source into the atmosphere provides for the large degree of control over the applied pulse characteristics; the propellant quantity controls the applied peak pressure, the venting of the high pressure gas into the source controls the pulse rise time, and the venting of the low pressure gas from the source to the atmosphere controls the pulse shape and decay time.

The propellant quantities for the SRISM tests will be designed to generate approximately 0.17, 0.34, 0.51, and 0.68 MPa load levels. The NTS test bed soil properties are similar to those described above. Three soil cantilever test bed dimensions will be tested as shown in Figure 8. Tests 1 and 2 will be performed on Soil Cantilever Dimension A which consists of a single 4.6-meter-long relief trench that is oriented parallel to the SRISM and 9.2 meters from the SRISM. This soil cantilever will then be modified for Tests 3 and 4 by adding two relief trenches that are 1.5 meters apart and oriented perpendicular to the SRISM (Soil Cantilever B). Finally, this soil cantilever will be further modified by placing a relief trench that is parallel to the SRISM and 4.6 m from the SRISM (Soil Cantilever C). The trench depths will be 4.6 m.

Figure 9 shows a plan view of the instrumentation layout for the SRISM evaluation tests. Four instrumentation bore holes, one in the free-field response region and three in the cantilever response region, will be used to place soil stress and triaxial accelerometer gages. Figure 10 shows a detail of the instrumentation bore hole design. Six soil stress gages will be placed at the SRISM-soil interface to measure the applied pressure distribution to the soil cantilever.

In addition to the active instrumentation, three single degree of freedom structures (SDOF) will be placed near the soil cantilever instrumentation bore holes. The motion of these SDOF structures will be recorded using high speed photography or accelerometer and used to validate the response spectra inferred by the acceleration measurements.

Future Plans

The results from the demonstration test will be used to validate the computer model used for predicting the response of the soil cantilever test bed at NTS. The validated computer model will be used to design the large-scale soil cantilever test bed described earlier. Construction of this test bed is planned to be completed by 1999. This state-of-the-art test facility will be used by a consortium of domestic and international researchers to enhance the structural integrity of buildings, bridges, power plants, and lifelines to earthquakes.

REFERENCES

1. Abrahamson, G. R., H. E. Lindberg, and J, R. Bruce, "Simulation of Strong Earthquake Motion with Explosive Line Source Array," Final Report prepared for the National Science Foundation, SRI Project 6004 (1977)

2. Bruce, J. R., H. E. Lindberg, and L. E. Schwer, "Soil Motion from Contained Explosions," Physical Modeling of Soil Dynamics Problems, Pre-Print 82-063, ASCE National Convention, Las Vegas, NV (1982).

3. Lindberg, H. E., R. Mak, and J. R. Bruce, "Calculation of Soil motion from Contained Explosive Arrays," Proceedings of the Eight World Conference on Earthquake Engineering, San Francisco, CA (1984).

4. Simons, J. W., A. N. Lin, and H. E. Lindberg, "Dynamic Testing of a Soil-Structure System Using the Technique of Repeatable Earth Shaking by Controlled Underground Expansion (RESCUE)," Final Report prepared for the National Science Foundation, SRI Project 4644 (1985).

5. Simons, J. W. and P. R. Gefken, "Improvements to the RESCUE Technique for Dynamic Testing of Soil-Structure Systems," 1989 ASME PVP Conference, Honolulu, HI (1989).

6. Gefken, P. R. and J. W. Simons, "Using the RESCUE Technique to Investigate the Soil-Structure Interaction for a Nuclear Reactor," Transactions of the 10th International Conference on Structural Mechanics in Reactor Technology, Anaheim, CA, 14-18 August 1989.

7. Simons, J. W. and P. R. Gefken, "Improvements to the RESCUE Technique for Dynamic Testing of Soil Structure Systems," Final Report prepared for the National Science Foundation, SRI Project 2134 (1990).

8. McCallen, D. "Assessment of a Technique for Simulating Earthquake Ground Motions with Controlled Explosions," Technical Report prepared for NeTI, Lawrence Livermore National Laboratory (1997).

9. Reinke, R. and G.Baladi, "Results of Seismic Refraction Surveys Conducted at the NeTI Earthquake Simulation Site, Yucca Flat, Area 9, NTS, February 1997.

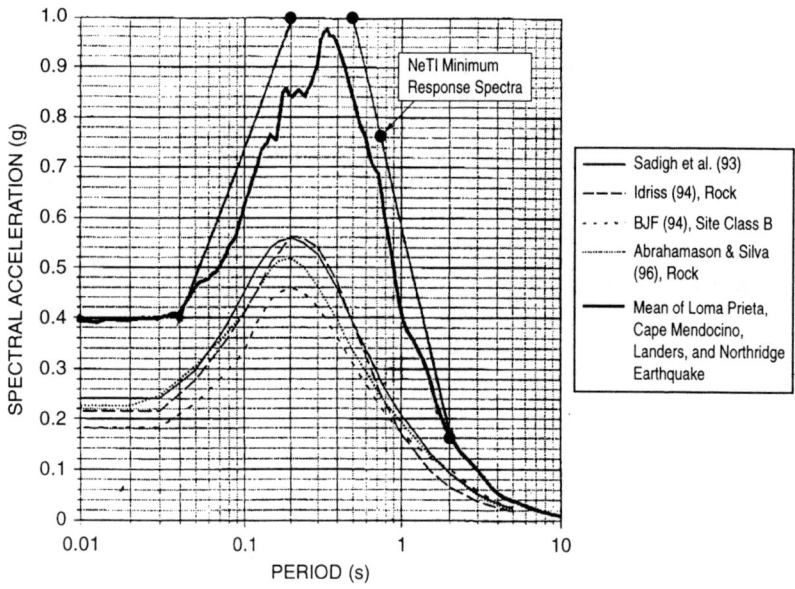

NAM-352525-9

Figure 1.   NeTI testing facility minimum response spectra.

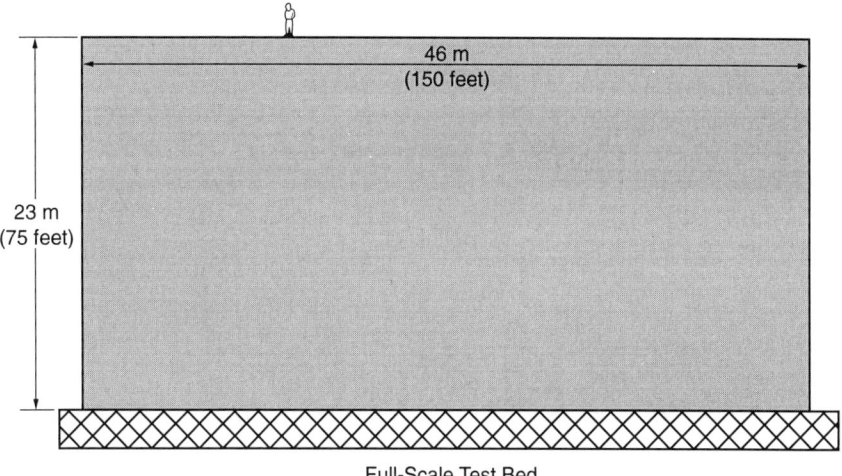

Full-Scale Test Bed

NM-353525-10

Figure 2.   NeTI soil cantilever test bed.

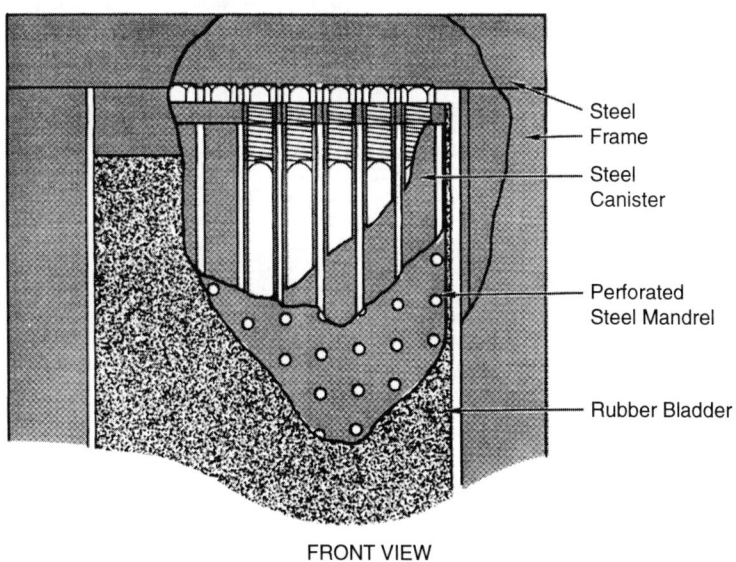

FRONT VIEW

RM-2134-38E

Figure 3.   SRI's RESCUE source design.

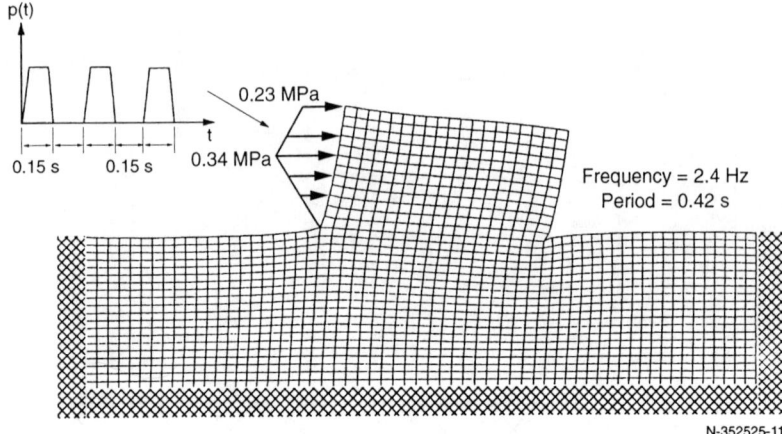

N-352525-11

Figure 4.   DYNA3D model of NeTI soil cantilever test bed.

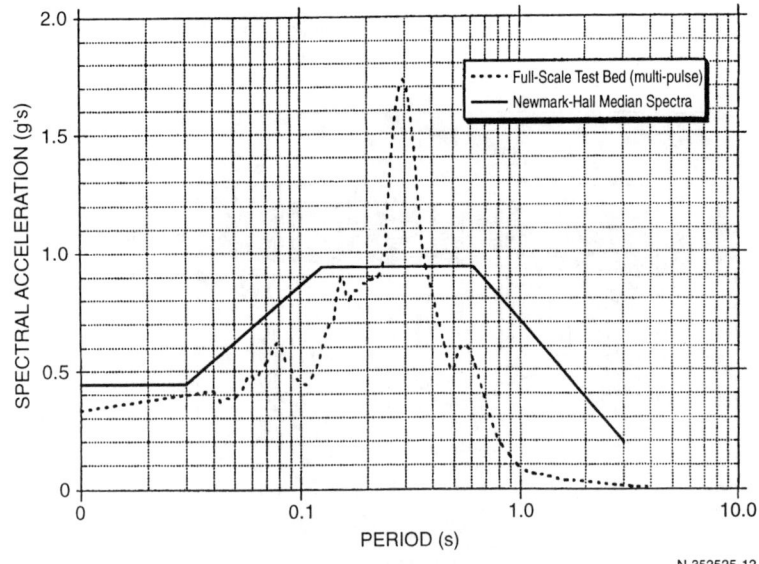

N-352525-12

Figure 5.   NeTI soil cantilever test bed response spectra.

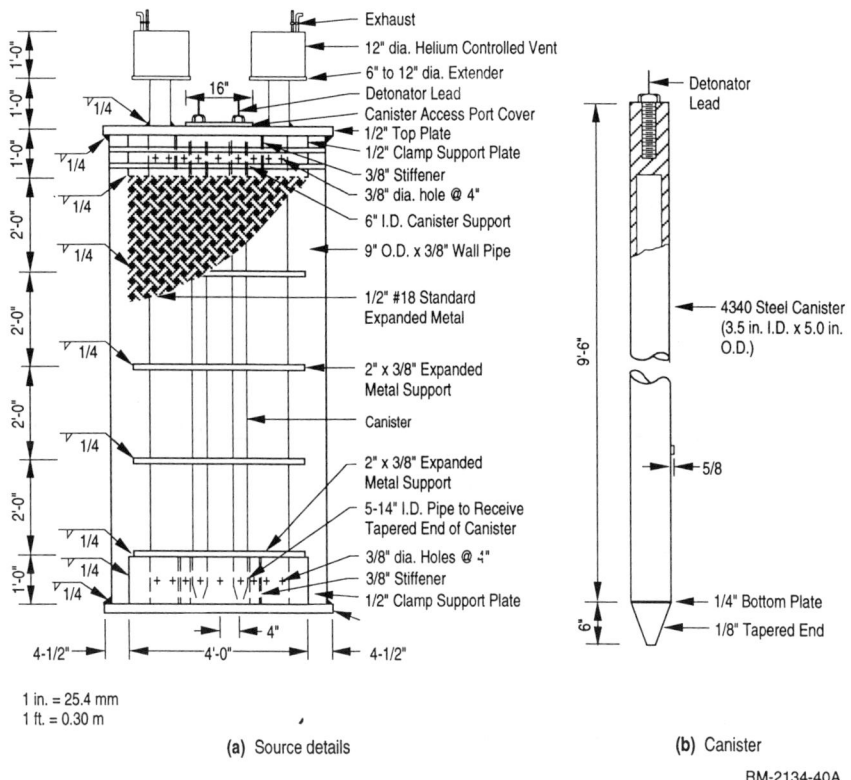

1 in. = 25.4 mm
1 ft. = 0.30 m

(a) Source details

(b) Canister

RM-2134-40A

Figure 6.   Plans for SRI Source Module (SRISM).

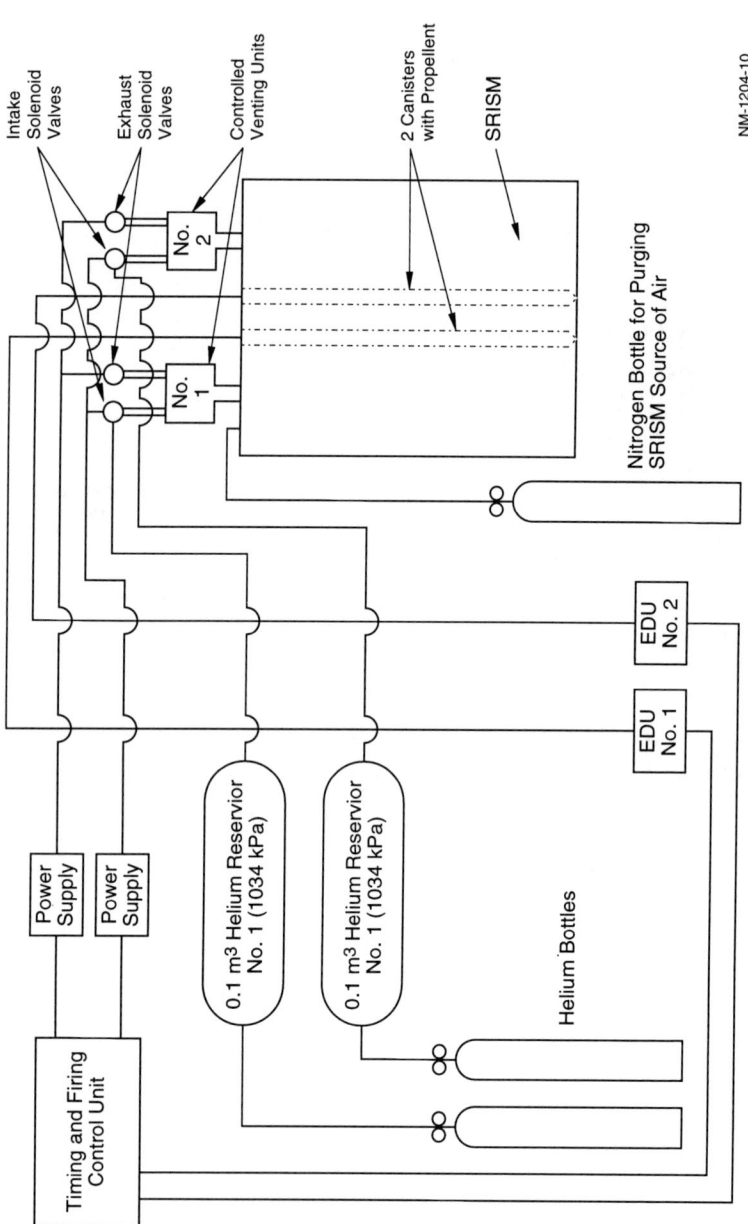

Figure 7.   SRISM hardware schematic diagram.

NM-1204-10

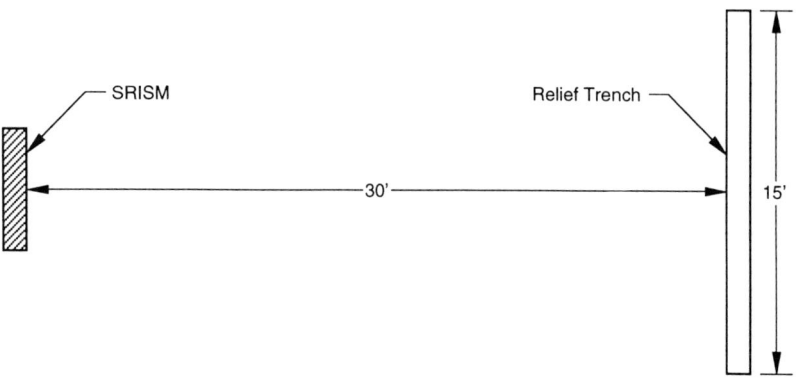

Soil Cantilever Dimension A

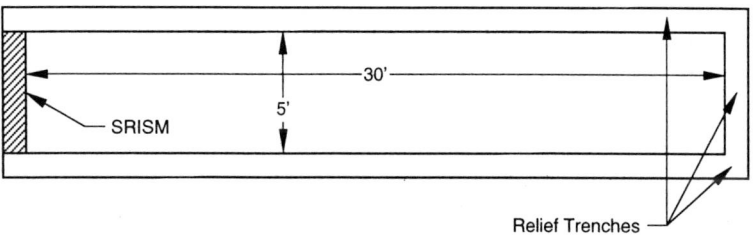

Soil Cantilever Dimension B

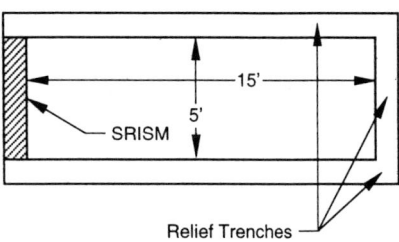

Soil Cantilever Dimension C

1 ft = 0.30 m

NM-1204-13

Figure 8. SRISM evaluation test configurations.

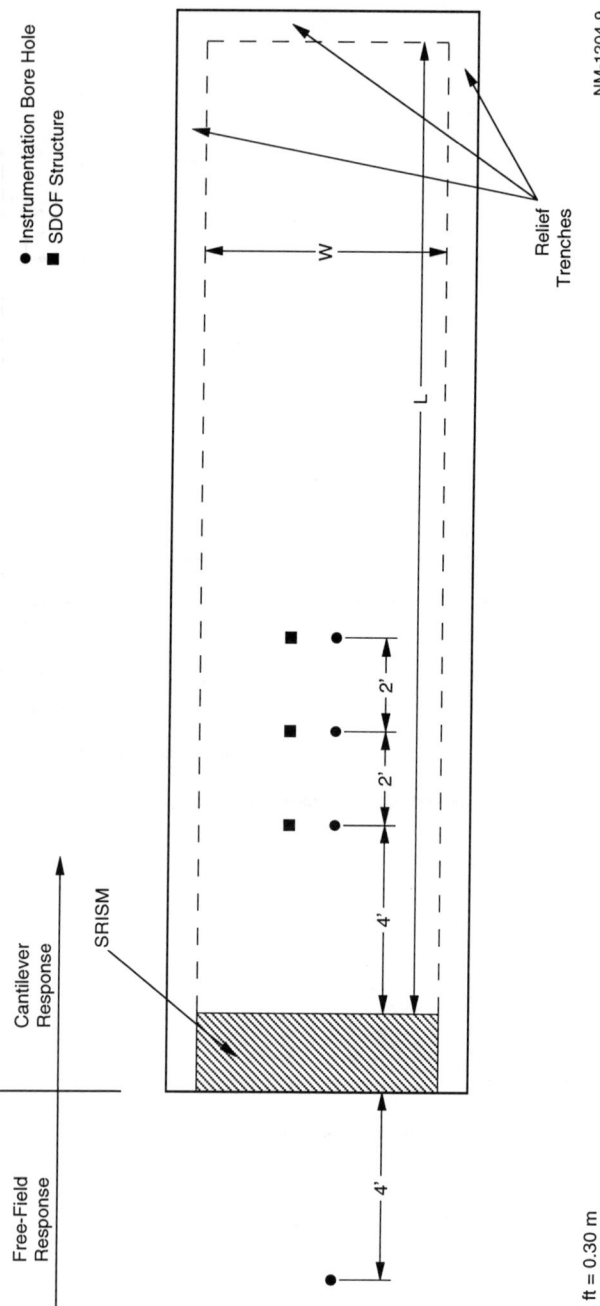

Figure 9.   Plan view of SRISM evaluation tests instrumentation arrangement.

1 ft = 0.30 m

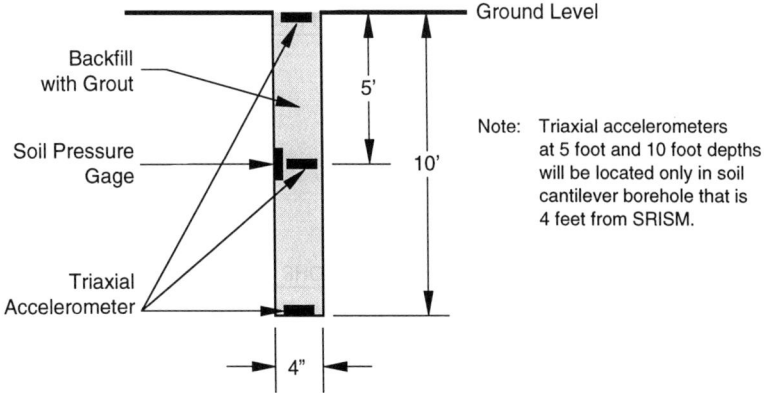

Instrumentation Bore Hole Detail

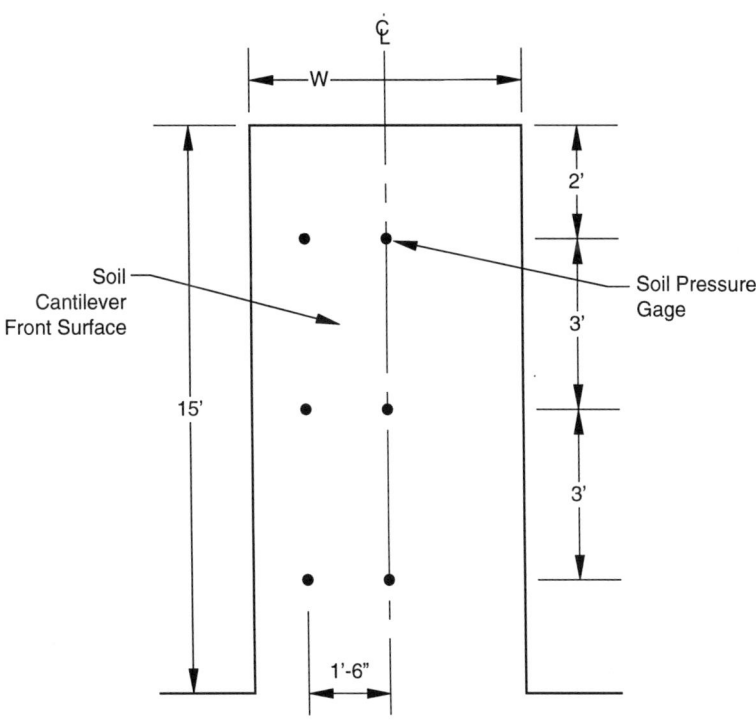

SRISM - Soil Interface Pressure Measurement Detail

1 in. = 25.4 mm
1 ft. = 0.30 m

NM-1204-12

Figure 10.   Instrumentation plan for SRISM evaluation tests.

Vibration Characteristics of Foundation Slabs
on a Two-Parameter Elastic Medium

Musharraf Zaman, Member[1]; M. Omar Faruque[2]; Adhir Agrawal[3];
Weng Low[3]; and Joakim G. Laguros, Fellow[4]

Abstract

Design of foundation slabs subjected to random and
cyclic loads requires an accurate assessment of the natural
frequencies and mode shapes. To this end, an improved
(finite element) formulation is presented in which the
exact solution for a beam resting on a Vlasov-type two-
parameter elastic medium is used to develop the shape
functions for the foundation slab element. A comprehensive
parametric study is conducted to investigate the effects of
geometric and material properties of the system and
boundary conditions. The results are presented in a non-
dimensional form that can be readily used for a wide range
of foundation slabs of practical significance.

Introduction

The analysis of free vibrations of foundation slabs
resting on an elastic medium (soil) is important to several
areas of geotechnical engineering such as highway and
airport pavements and mat foundations. Design of
foundations subjected to random loads from earthquakes and
cyclic loads from operating machines and traversing traffic

---

[1]Professor; [2]Former Visiting Assistant Professor; [3]Former
Graduate Student; and [4]Professor Emeritus; University of
Oklahoma; School of Civil Engineering and Environmental
Science; 202 West Boyd Street, Room 334; Norman, OK 73019

requires a realistic evaluation of the natural frequencies and mode shapes.

To analyze structural foundations as beams or plates resting on an elastic medium and subjected to static and dynamic loads, several approaches have been suggested and reported in the literature (Kerr, 1964; Yang, 1972; Selvadurai, 1979; Nogami and Lam, 1987; Zaman et al, 1993, 1995). Because an analytical or a closed form solution of the governing differential equations associated with such problems is extremely difficult, if not impossible (Dawe and Roufaeil, 1980), different numerical solution techniques have been attempted including the finite difference, the finite element, the boundary integral equation, and the energy methods. To render the analysis simpler various assumptions are made among which treating the supporting soil as a Winkler medium is probably the most significant. To circumvent the Winkler limitations of shear, two-parameter models were introduced to account for the shear interactions between discrete Winkler springs (Nogami and Lam, 1980; Selvadurai, 1979). Alternatively, the soil medium can be idealized as two- or three-dimensional solid elements in a numerical analysis, but it would significantly increase the computational effort, particularly for dynamic problems, and it may not be justified on a time-cost basis for some applications (Zaman and Alvappillai, 1995).

A number of two-parameter models are available in the literature; among these, the models proposed by Filonenko-Borodich (1940), Hetenyi (1946), Pasternak (1954), and Vlasov-Leontiev (1966) are some of the particularly notable ones. In these models, except the Vlasov model, the pure shear deformation between the individual membrane elements, elastic plate elements or elastic layers are capable of undergoing shear interaction. The Vlasov model, on the other hand, achieves the shear interaction by imposing restrictions on the distribution of displacements and stresses in the elastic halfspace (Kerr, 1964).

Objective

In the present study, the free vibration response of foundation slabs is investigated within the framework of

the finite element (FE) method. The supporting elastic
soil medium is idealized by the Vlasov two-parameter model.
The dynamic soil-structure interaction effects are included
by considering the foundation slab and the supporting two-
parameter elastic medium simultaneously in the governing
differential equation (GDE). The primary contribution of
the proposed FE algorithm lies in the development of the
shape functions that are consistent with the foundation
slab displacement profile. In this study, Bogner's
approach (1966) of using a cubic Hermite polynomial is used
to evaluate the desired stiffness and mass matrices for the
thin plate (foundation slab) elements. However, instead of
using the Hermite polynomial to represent displacements,
the shape functions derived by Low (1990) are employed.
Low's functions were derived from the actual deflected
shape of a beam resting on the Vlasov's two-parameter
elastic medium. These stiffness and mass matrices are
utilized to develop a FE algorithm for the free vibration
analysis of rectangular and square foundation slabs. A
parametric study is conducted to investigate the effect of
various parameters on the natural frequencies and mode
shapes of foundation slabs. Damping and forced vibration
aspects were not addressed at this stage.

Finite element formulation

The free vibration analysis of a foundation slab
resting on a two-parameter elastic medium using the finite
element (FE) technique requires formulation of stiffness
and mass matrices of the foundation slab-elastic medium
system. Bogner (1966) used a cubic Hermite polynomial to
model the transverse displacement of a beam element (Desai
and Abel, 1972). In the present study, the basic function
for a beam is used to model the transverse displacement of
a foundation slab element. The foundation slab is
discretized by four-noded quadrilateral elements (Figure 1)
having the following degrees of freedom at each node:
vertical displacement, w_i; rotation with respect to x-axis,
$\theta_{xi} = \partial w/\partial x$; rotation with respect to y-axis, $\theta_{yi} = \partial w/\partial y$; and
torsion $\theta_{xyi} = \partial^2 w/(\partial x\ \partial y)$, where subscript i represents the
node number.

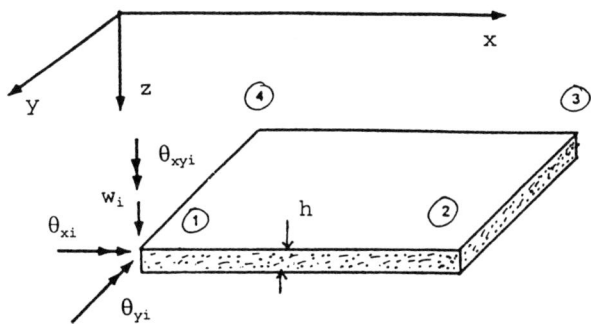

Figure 1. A Four-Noded Foundation Element With Degrees of Freedom Shown at One Node

This results in a total of sixteen degrees-of-freedom per element. As shown by Bogner (1966), the Hermite polynomials that can be used as shape functions to represent the displacement characteristics of a one-dimensional beam in x and y directions can be expressed as:

$$
\left.
\begin{aligned}
NX1 &= 1-3s^2+2s^3 \\
NX2 &= s^2(3-2s) \\
NX3 &= as(s-1)^2 \\
NX4 &= as^2(s-1)
\end{aligned}
\right\}
\quad
\begin{aligned}
& s = x/a \\
& 0 \le s \le 1 \\
& \frac{\partial}{\partial x} = \frac{1}{a}\frac{\partial}{\partial s}
\end{aligned}
\tag{1}
$$

$$
\left.
\begin{aligned}
NY1 &= 1-3t^2+2t^3 \\
NY2 &= bt(3-2t) \\
NY3 &= t^2(t-1)^2 \\
NY4 &= bt^2(t-1)
\end{aligned}
\right\}
\quad
\begin{aligned}
& t = y/b \\
& 0 \le t \le 1 \\
& \frac{\partial}{\partial y} = \frac{1}{b}\frac{\partial}{\partial t}
\end{aligned}
\tag{2}
$$

where, a and b are the linear dimensions of a foundation slab element in the x and y directions, respectively.

By using these shape functions, the displacement field for a foundation slab element subjected to bending can be written as (Desai and Abel, 1972):

$$
\begin{aligned}
w(x,y) = {}& N1w_1 + N2\,\theta_{x1} + N3\,\theta_{y1} + N4\,\theta_{xy1} \\
& + N5w_2 + N6\,\theta_{x2} + N7\,\theta_{y2} + N8\,\theta_{xy2} \\
& + N9w_3 + N10\,\theta_{x3} + N11\,\theta_{y3} + N12\,\theta_{xy3} \\
& + N13w_4 + N14\,\theta_{x4} + N15\,\theta_{y4} + N16\,\theta_{xy4}
\end{aligned}
\tag{3}
$$

where N1, N2,...N16 are functions of NX1, NX2,...NY4 (Desai and Abel, 1972).

In the present study, instead of using NX1, NX2,...NY4 that are applicable to a purely beam bending problem, a different set of shape functions is utilized that reflect the actual response of a beam resting on a Vlasov-Leontiev (1966) type two-parameter elastic medium and subjected to bending.

To this end, consider the GDE for bending of a beam supported on a two-parameter elastic medium:

$$EI\frac{d^4w(x)}{dx^4} - k_1\frac{d^2w(x)}{dx^2} + k_2w(x) = 0 \tag{4}$$

where EI is the flexural rigidity, $w(x)$ the deflection at a point x, E the modulus of elasticity (beam), I the moment of inertia of the beam cross section, k_1 the moment foundation modulus, and k_2, the Winkler or subgrade modulus.

As shown by Selvadurai (1979) and Low (1990), by employing the method of initial parameters, the general solution of Eq. (4) can be written as:

$$w(x) = A_1w_0 + A_2\theta_0 + A_3M_0 + A_4N_0 \tag{5}$$

where,

$$A_1 = \phi_2 - \frac{\mu^2-\alpha^2}{2\mu\alpha}; \quad A_2 = \frac{1}{2\lambda}\left[\frac{\phi_1}{\mu} + \frac{\phi_3}{\alpha}\right]; \quad A_3 = -\frac{\phi_4}{2\mu\alpha\lambda^2 EI}; \quad A_4 = \frac{-1}{4\lambda^3 EI}\left[\frac{\phi_1}{\mu} - \frac{\phi_3}{\alpha}\right];$$

$$\phi_1 = \cos(\alpha\lambda x)\sinh(\mu\lambda x); \quad \phi_2 = \cos(\alpha\lambda x)\cosh(\mu\lambda x); \quad \phi_3 = \sin(\alpha\lambda x)\cosh(\mu\lambda x);$$

$$\phi_4 = \sin(\alpha\lambda x)\sinh(\mu\lambda x); \quad \lambda = \left[\frac{k_2}{4EI}\right]^{1/4}; \quad \mu = \left[1 + \frac{k_1}{k_2}\lambda^2\right]^{1/2} \text{ for } 0 \le \mu \le 1;$$

$$\alpha = \left[1 - \frac{k_1}{k_2}\lambda^2\right]^{1/2} \text{ for } 0 \le \alpha \le 1; \quad k_1 = \frac{E_o b[s\ \sinh(2\gamma H/s) - 2\gamma H]}{8\gamma(1+v_o)\sinh^2(\gamma H/s)};$$

$$k_2 = \frac{E_o b\gamma[s\ \sinh(2\gamma H/s) + 2\gamma H]}{4s^2(1-v_o^2)\sinh^2(\gamma H/s)}; \quad E_o = \frac{E_s}{(1-v_s^2)}; \quad v_o = \frac{v_s}{(1-v_s)}$$

Here w_0, θ_0, M_0, and N_0 are the deflection, slope, bending moment and transverse shear force, respectively, at the left end of the beam (at $x = o$). Also, E_s and v_s are the elastic properties of soil and H is the thickness of the

homogeneous soil layer and s is the half-width of the beam.
As shown by Low (1990), in view of Eqs. (4) and (5) the
shape functions N_1, N_2, N_3, and N_4 for the beam element can
be expressed in a compact form as:

$$N_1 = \frac{1}{2\mu\lambda}\left[2\mu\alpha\phi_2 - (\mu^2 - \alpha^2)\phi_4\right] - \frac{\phi_4}{2\mu\alpha\lambda^2 EI}k_{21} + \frac{1}{4\mu\alpha\lambda^3 EI}\left[\mu\phi_3 - \alpha\phi_1\right]k_{11} \tag{6a}$$

$$N_2 = \frac{1}{2\mu\alpha\lambda}\left[\alpha\phi_1 + \mu\phi_3\right] - \frac{\phi_4}{2\mu\alpha\lambda^2 EI}k_{22} + \frac{1}{4\mu\alpha\lambda^3 EI}\left[\mu\phi_3 - \alpha\phi_1\right]k_{12} \tag{6b}$$

$$N_3 = -\frac{\phi_4}{2\mu\alpha\lambda^2 EI}k_{23} + \frac{1}{4\mu\alpha\lambda^3 EI}\left[\mu\phi_3 - \alpha\phi_1\right]k_{13} \tag{6c}$$

$$N_4 = -\frac{\phi_4}{2\mu\alpha\lambda^2 EI}k_{24} + \frac{1}{4\mu\alpha\lambda^3 EI}\left[\mu\phi_3 - \alpha\phi_1\right]k_{14} \tag{6d}$$

where

$$k_{11} = \frac{\mu\alpha k_2}{\lambda}\left[\frac{\mu\sin(\alpha\lambda L)\cos(\alpha\lambda L) + \alpha\sinh(\mu\lambda L)\cosh(\mu\lambda L)}{F}\right] = k_{33} \tag{6e}$$

$$k_{21} = \frac{k_2}{2\lambda^2}\left[\frac{\mu^2\sin^2(\alpha\lambda L) + \alpha^2\sinh^2(\mu\lambda L)}{F}\right] = k_{12} = -k_{34} = -k_{43} \tag{6f}$$

$$k_{31} = -\frac{\mu\alpha k_2}{\lambda}\left[\frac{\mu\sin(\alpha\lambda L)\cosh(\mu\lambda L) + \alpha\cos(\alpha\lambda L)\sinh(\mu\lambda L)}{F}\right] = k_{13} \tag{6g}$$

$$k_{41} = \frac{\mu\alpha k_2}{\lambda^2}\left[\frac{\sin(\alpha\lambda L)\sinh(\mu\lambda L)}{F}\right] = k_{14} = -k_{23} = -k_{32} \tag{6h}$$

$$k_{22} = \frac{\mu\alpha k_2}{2\lambda^3}\left[\frac{\alpha\sinh(\mu\lambda L)\cosh(\mu\lambda L) - \mu\sin(\alpha\lambda L)\cos(\alpha\lambda L)}{F}\right] = k_{44} \tag{6i}$$

$$k_{42} = \frac{\mu\alpha k_2}{2\lambda^3}\left[\frac{\mu\sin(\alpha\lambda L)\cosh(\mu\lambda L) - \alpha\cos(\alpha\lambda L)\sinh(\mu\lambda L)}{F}\right] = k_{24} \tag{6j}$$

and

$$F = \alpha^2\sinh^2(\mu\lambda L) - \mu^2\sin^2(\alpha\lambda L) \tag{6k}$$

These shape functions, called "basic functions," are used
to derive the components $(N_1, N_2, \ldots N_{16})$ in Eq. (3) that

represent the displacement function of a rectangular (foundation slab) plate element, following the approach suggested by Bogner (1966). Further details of the shape functions and expressions for N_1, N_2, ...N_{16} are given by Agrawal (1993).

Stiffness matrices

By considering the GDE for a foundation slab element resting on a two-parameter elastic medium, the total strain energy of the element and its variation, the bending stiffness matrix $[k_p]$ can be expressed as:

$$[k]_p = h \int_o^x \int_o^y [B]_p^T [C_p][B]_p \ a \ b \ dxdy \tag{7}$$

where $[B]_p$ is a transformation matrix, $[C_p]$ is the elastic constitutive relation matrix, h is the foundation slab thickness and T represents transpose. For the foundation slab element under consideration, the $[B]_p$ and $[C_p]$ matrices are given by:

$$[B]_p = \begin{bmatrix} \dfrac{\partial^2 N1}{\partial x^2}, \dfrac{\partial^2 N2}{\partial x^2}, \ldots, \dfrac{\partial^2 N16}{\partial x^2} \\ \dfrac{\partial^2 N1}{\partial y^2}, \dfrac{\partial^2 N2}{\partial y^2}, \ldots, \dfrac{\partial^2 N16}{\partial y^2} \\ \dfrac{\partial^2 N1}{\partial x \partial y}, \dfrac{\partial^2 N2}{\partial x \partial y}, \ldots, \dfrac{\partial^2 N16}{\partial x \partial y} \end{bmatrix} \tag{8}$$

$$[C_p] = \frac{Eh^3}{12(1-v^2)} \begin{bmatrix} 1 & v & 0 \\ v & 1 & 0 \\ 0 & 0 & \dfrac{(1-v)}{2} \end{bmatrix} \tag{9}$$

where E and v are the elastic properties of the foundation slab material. Likewise, the stiffness matrix of the two-parameter Vlasov elastic medium element $[k]_f$, supporting the foundation slab element can be expressed as:

$$[k]_f = [k]_1 + [k]_2 \tag{10}$$

where,

$$[k]_1 = \int_o^x \int_o^y [N]^T k_1 [N] \ a \ b \ dxdy \tag{11a}$$

$$[k]_2 = 2 \int_o^x \int_o^y [B]_f^T k_2 [B]_f \, a \, b \, dxdy \qquad \text{(11b)}$$

and

$$[B]_f = \begin{bmatrix} \dfrac{\partial N1}{\partial x}, & \dfrac{\partial N2}{\partial x}, & \cdots\cdots, & \dfrac{\partial N16}{\partial x} \\[2mm] \dfrac{\partial N1}{\partial y}, & \dfrac{\partial N2}{\partial y}, & \cdots\cdots, & \dfrac{\partial N16}{\partial y} \end{bmatrix} \qquad \text{(11c)}$$

Mass matrix of the foundation slab element

The mass of the supporting soil medium is neglected here for simplicity, and only the foundation slab mass is used in evaluating the element mass matrix. The consistent mass formulation is employed. Accordingly, the element mass matrix is given as:

$$[m] = \int_v \rho [N]^T [N] \, dv \qquad \text{(12)}$$

where ρ is the mass density of the foundation slab material and V is the volume of the foundation slab element.

Modal analysis

By applying the principle of virtual work and the variational principle, the governing differential equation for a foundation slab, which is modeled as a thin plate resting on a two-parameter elastic medium, can be written in the form:

$$[K] \{W\} + [M] \{\ddot{w}\} = \{0\} \qquad \text{(13)}$$

where $[K]$ and $[M]$ are the system stiffness and system mass matrices, respectively, obtained by assembling the element stiffness and mass matrices.

The generalized eigenproblem for this case is indicated as:

$$[K] \{\phi\} = \{w^2\} [M] \{\phi\} \qquad \text{(14)}$$

from which the eigenvector or the mode shapes $\{\phi\}$ and the corresponding eigenvalues w or the natural frequencies of the foundation slab-elastic support system are determined.

Numerical results

In most numerical analyses only the first few natural frequencies of vibration and the corresponding mode shapes are important from an engineering design point of view. Therefore, natural frequencies and modes shapes are determined here only for the first five modes of free vibration of a foundation slab resting on a two-parameter elastic medium. These frequencies represent the lowest five frequencies of the foundation slab. A comprehensive parametric study was conducted to investigate the influence of various factors. From the comprehensive results reported by Agrawal (1993) only a few selected ones are presented herein due to space limitation.

Comparison with exact solutions

From Table 1 it is evident that the exact solutions for a simply supported foundation slab or plate (without soil support) and the numerical results obtained from the classical plate analyses (Gorman, 1982) agree well with the results obtained from the present finite element (FE) analysis. The geometric and material properties used by Gorman (1982) are also used here. The minor differences in the frequencies are due to the coarseness of the finite element mesh (16 elements) used to discretize the plate.

MODE NO.	EXACT ANALYSIS (GORMAN, 1982)	BOGNER'S ANALYSIS (BOGNER, 1966)	F. E. ANALYSIS	PERCENT DIFFERENCE
1	1035.0	1069.0	1039.9	0.47
2,3	2587.0	2669.0	2606.0	0.74
4	4138.0	4337.0	4168.1	0.73
5,6	5173.0	5368.0	5297.8	2.06

Table 1. Comparison of Frequencies for a Simply Supported Square Plate Without Any Soil Support

As shown by Agrawal (1993), the computed natural frequencies showed a significant improvement when 36 elements were used to discretize the plate instead of 16 elements. However, the computed frequencies did not exhibit the same degree of improvement when the number of elements was increased from 36 to 64.

Non-dimensional parameters

For convenience, numerical results are presented here in a non-dimensional form. Two non-dimensional parameters, flexibility index (γ) and frequency parameter (Ω), are defined for this purpose as follows:

$$\gamma = \frac{\pi E_o L^2 B}{8 D (1 - v_o^2)} \tag{15a}$$

$$\Omega = w \sqrt{\frac{(1+v)\rho LB}{E}} \tag{15b}$$

where L = length, B = width, E = modulus, ν = Poission's ratio, D = natural frequency, w = flexural and rigidity. Yang (1972) and Dawe and Roufaeil (1980) used similar non-dimensional parameters. A third parameter, κ, which is defined as the ratio between k_1 and k_2, is used to investigate the effect of supporting elastic medium. k_1 and k_2 are related to the conventional elastic properties of the soil medium as follows:

$$k_1 = E_o \psi / [B(1 - v_o^2)] \quad \text{and} \quad k_2 = E_o B / [16 \psi (1 + v_o)].$$

Following Selvadurai (1979), the parameter Ψ is assumed as 1.5 in this study.

Effect of κ on γ and Ω

Four different values of κ, namely 0, 50, 100 and 200, are selected to study the variations in γ and Ω with κ. Note that κ = 0 represents a foundation slab resting on the Winkler medium. Two different types of boundary conditions are considered: a simply supported foundation slab at all four edges and a clamped foundation slab at all four edges.

Figure 2 shows the relationship between Ω and γ for a square foundation slab (φ = L/B = 1) resting on a Winkler medium (κ = 0). A similar relationship for the two-parameter medium with κ = 100 is shown in Figure 3, for the first five modes. Similar graphical results for a rectangular foundation slab (φ = L/B = 2) for κ = 0 and 100 are shown in Figures 4 and 5, respectively. Overall, it is observed that the variation in Ω with γ is less sensitive at lower modes than at higher modes. This is true for both square and rectangular slabs. For a given flexibility index (γ), the frequency parameter (Ω) for a square foundation slab is higher than that for a rectangular foundation slab. Such differences are particularly noticeable for higher modes of vibration. One of the advantages of using non-dimensional parameters for presentation of results is that these results can be used for other slabs by assigning appropriate geometric and material properties.

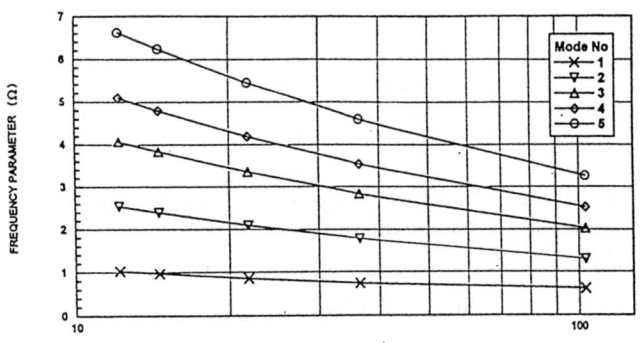

Figure 2. Variation of Flexibility Index Vs Frequency Parameter for a Plate Simply Supported at all Four Edges. [κ = 0; φ = 1.0]

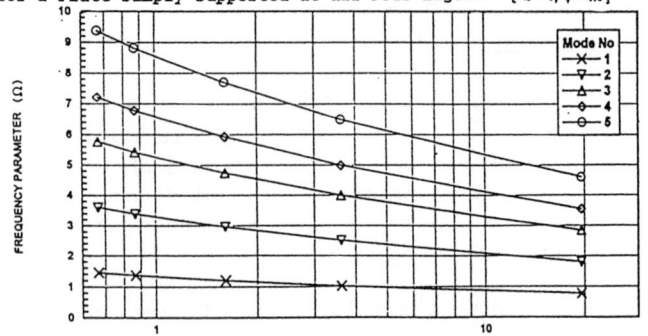

Figure 3. Variation of Flexibility Index Vs Frequency Parameter for a Plate Simply Supported at all Four Edges. [κ = 100; φ = 1.0]

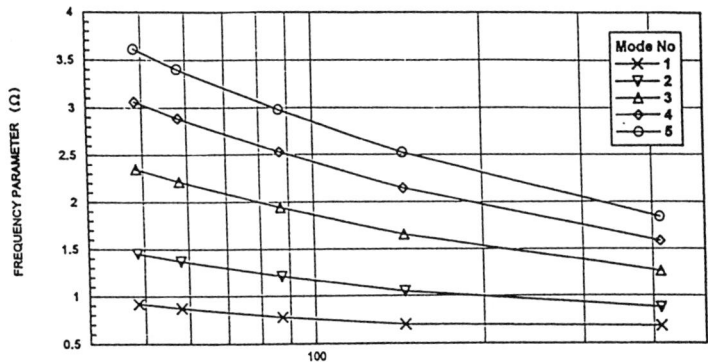

Figure 4. Variation of Flexibility Index Vs Frequency Parameter for a Plate Simply Supported at all Four Edges. [κ = 0; φ = 2.0]

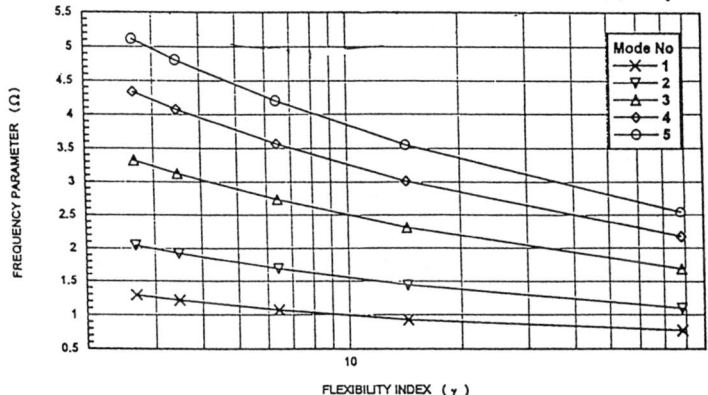

Figure 5. Variation of Flexibility Index Vs Frequency Parameter for a Plate Simply Supported at all Four Edges. [κ = 100; φ = 2.0]

Effect of boundary conditions

To demonstrate the effect of boundary conditions, the frequency parameter (Ω) values for a clamped square foundation slab are plotted as a function of γ in Figure 6. Comparing Figures 3 and 6, it is observed that Ω for a clamped foundation is consistently higher than that of a simply supported foundation slab, as expected. However, as noted by Agrawal (1993), the values of the flexibility index for clamped foundation slabs are lower than for simply supported foundations (keeping other factors unchanged) for all modes of vibration. A decrease in the value of the flexibility index can be attributed to constraining of more degrees-of-freedom in the clamped foundation as compared to the simply supported foundation slab. Also, the clamped foundation slab shows increased stiffness and frequency parameters compared with simply supported slabs. An increase in the frequency parameters at higher modes is also expected.

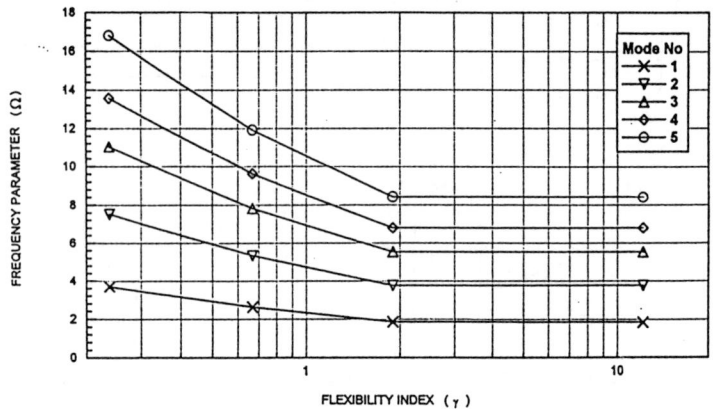

Figure 6. Variation of Flexibility Index Vs Frequency Parameter for a Plate Clamped at all Four Edges. [μ = 0.20; φ = 1.0]

Effect of loss of support

Due to the vibration of a foundation, the loss of contact between a foundation slab and supporting medium may occur. To simulate this effect, partially supported slabs are analyzed and the resulting mode shapes are compared with those for the fully unsupported slabs.

The results for the mode shapes are compared for the various values of β (ratio of the supported length to the total length of foundation slab) with the corresponding mode shapes of a fully supported slab along with supports at all four edges. The first mode of vibration of a fully unsupported foundation slab ($\beta = 0.0$) in Figure 7 is compared with a partially supported foundation slab ($\beta = 0.25$) in Figure 8. Because of the elastic medium support at quarter length of the foundation slab near the right edge (Figure 8), the foundation slab is stiffer than a

fully unsupported foundation slab which causes a decrease
in the normalized mode shape amplitude near the right edge
of the partially unsupported foundation slab. Contours of
normalized mode shape amplitude in Figure 8 show a shift
toward the left edge because of the increased stiffness of
the foundation slab near the right edge.

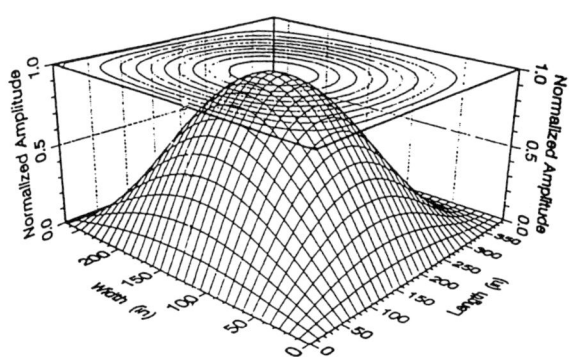

Figure 7. Mode Shape of Free Vibration of a Plate Simply
Supported at all Four Edges:1st Mode [μ=.45; φ=1.5; β=0.0]

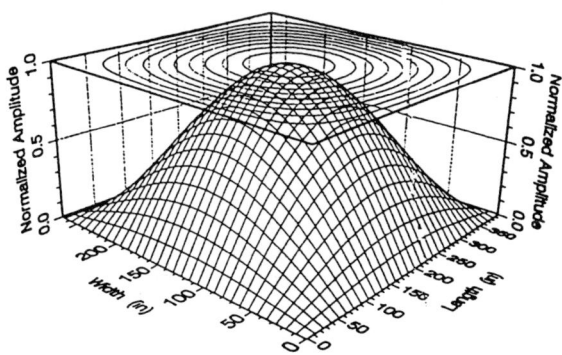

Figure 8. Mode Shape of Free Vibration of a Plate Simply
Supported at all Four Edges:1st Mode [μ=.45; φ=1.5; β=0.25]

Concluding remarks

The proposed analysis of a foundation slab resting on a two-parameter elastic medium can be used to evaluate the vibration characteristics of a wide range of foundation slabs used in engineering applications.

The flexibility index significantly affects the free vibration characteristics of a slab; the natural frequency decreases with an increase in the flexibility index. This can be achieved by increasing the length and width of a foundation slab, reducing foundation thickness, and decreasing the Poisson's ratio of the supporting elastic medium. An increase in the ratio of $\kappa = k2/k1$ can significantly affect the frequency parameter, natural frequency and flexibility index γ. An increase in the ratio κ causes an increase in the flexibility index and decrease the frequency parameter and natural frequencies of a foundation slab resting on a two-parameter medium. Also, a loss of elastic support has profound effect on mode shapes of the slab-elastic support system.

APPENDIX I.   REFERENCES

Agrawal, A.   *Free Vibration Analysis of Rectangular Plates Resting on a Two-Parameter Elastic Medium*, M.S. thesis, the University of Oklahoma, OK, 1993.

Bogner, F.K., Fox, R.L., and Schmit, L.A.   The generation of inter-element-compatible stiffness and mass matrices by the use of interpolation formulas, *Proceedings of the Conference on Matrix Methods in Structural Mechanics*, AFFKL-TR-66-80, pp. 397-440, Wright-Patterson Air Force Base, OH, 1966.

Dawe, D.J. and Roufaeil, O.L.   "Vibration Analysis of Rectangular Mindlin Plates by Finite Strip Method," Computers and Structures, Vol. 12, pp. 833-843, 1980.

Desai, C.S. and Abel, J.F.   *Introduction to the Finite Element Method*, Von Nostrand Reinhold, New York, NY, 1972.

Filonenko-Bordich, M.M.   Some approximate theories of the elastic foundation, (in Russian), *Uchenyie Zapiski Moskovskogo Gosudarstuennogo Universiteta Mechanika*, 1940, No. 46, pp. 3-18.

Gorman, D.J. Free vibration analysis of rectangular plates, Elsevier North Holland, New York, 1982.

Hetenyi, M. *Beams on Elastic Foundation*, University of Michigan Press, Ann Arbor, MI, 1946.

Kerr, A.D. Elastic and viscoelastic foundation models, *Journal of Applied Mechanics*, Transaction of the ASME, pp. 491-498, 1964.

Low, W.C. *Analysis of Free Vibration of Beams and Plane Frame Resting on the Vlasov's Two-Parameter Elastic Medium*, M.S. thesis, University of Oklahoma, OK, 1990.

Nogami, T. and Lam, Y.C., "Two-Parameter Layer Model for Analysis of Slab on Elastic Foundation," *J. of Engineering Mechanics*, ASCE, 1987, Vol. 113, No. 9, pp. 1279-1291.

Pasternak, P.L. On a new method of analysis of an elastic foundation by means of two foundation constants, (in Russian), *Gosudarstvennoe Izdatelstvo Literaturi po Stroitielstvu I Arkhitekture*, Moscow, 1954.

Selvadurai, A.P.S. *Elastic Analysis of Soil-Foundation Interaction*, Elsevier Scientific Publishing Company, Amsterdam, Holland, 1979.

Vlasov, V.Z. and Leontev, M.M. *Beams, Plates and Shells on Elastic Foundation*, Published for NASA and NSF by the Israel Program for Scientific Translation, 1966.

Yang, T.Y. "A Finite Element Analysis of Plates on a Two-Parameter Foundation Model," <u>Computers & Structures</u>, Vol. 2, pp. 593-614, 1972.

Zaman, M.M., Alvappillai, A., and Taheri, M.R. Dynamic analysis of concrete pavements resting on a two-parameter medium, *International Journal for Numerical Methods in Engineering*, 1993, Vol. 36, pp. 1465-1486.

Zaman, M.M. and Alvappillai, A. Contact Element Model for Dynamic Analysis of Jointed Concrete Pavements. *J. Of Transp. Eng.*, ASCE, 1995, Vol. 121, No. 5, pp. 425-433.

SUBJECT INDEX

Page number refers to the first page of paper

Boundary conditions, 26, 122

Caissons, 11
Centrifuge, 47
Centrifuge model, 76
Computer programs, 76
Concrete piles, 76
Cyclic loads, 122

Design, 122
Displacement, 91
Displacements, 64
Dynamic response, 1, 26, 64
Dynamic tests, 47

Earthquake damage, 76
Earthquakes, 47, 91, 107
Elastic media, 122

Footings, 1
Foundations, 26, 122

Ground motion, 107

In situ tests, 1

Layered soils, 64
Liquefaction, 11, 47, 76

Model tests, 91
Modeling, 47
Models, 26, 64

Numerical models, 76

Pile foundations, 64
Pile settlement, 64
Pile structures, 47
Pore water pressure, 11
Profiles, 1

Quays, 11

Seismic response, 47, 107
Shake table tests, 91
Sheet piles, 11
Simulation, 91, 107
Slabs, 122
Soil dynamics, 1, 26
Soil-pile interaction, 47, 76
Soil-structure interaction, 1, 11, 26, 91, 107
Stiffness, 1

Testing, 107

Vibration, 122

Walls, 11

AUTHOR INDEX

Page number refers to the first page of paper

Abdoun, T., 76
Abghari, Abbas, 47
Agrawal, Adhir, 122
Arango, Ignacio, 107

Boulanger, Ross W., 47

Curran, Don, 107

Dobry, R., 76

Faruque, M. Omar, 122

Gefken, Paul, 107
Gurbuz, Orhan, 107
Guzina, B. B., 1

Iai, Susumu, 11
Ichii, Koji, 11

Konagai, Kazuo, 26, 91
Kutter, Bruce L., 47

Laguros, Joakim G., 122
Low, Weng, 122

McCallen, Dave, 107
Mikami, Atsushi, 26
Mote, Pete, 107

Nogami, Toyoaki, 26, 91

O'Rourke, T. D., 76

Pak, R. Y. S., 1

Vallabhan, C. V. Girija, 64

Wilson, Daniel W., 47

Zaman, Musharraf, 122
Zhen, Shengli, 26

NEW GEOTECH BOOKS FROM GEOLOGAN
JULY 15-19, 1997 • LOGAN, UTAH

Dredging and Management of Dredged Material
● Jay N. Meegoda, Thomas H. Wakeman III, Arul K. Arulmoli, and William J. Librizzi, Editors
Geotechnical Special Publication #65
Proceedings of three sessions held in conjunction with Geo-Logan sponsored by the Soil Properties Committee of The Geo-Institute of the ASCE
208 pp, List, $25.00; ASCE member, **$18.75**
(#40254)

Ground Improvement, Ground Reinforcement, Ground Treatment: Developments 1987-1997
● Vernon R. Schaefer, Editor
Geotechnical Special Publication #69
Proceedings of sessions sponsored by the Committee on Soil Improvement and Geosynthetics of The Geo-Institute of the ASCE in conjunction with Geo-Logan '97
632 pp, List, $53.00; ASCE member, **$39.75**
(#40260)

Grouting: Compaction, Remediation and Testing
● C. Vipulanandan, Editor
Geotechnical Special Publication #66
Proceedings of sessions sponsored by the Grouting Committee of The Geo-Institute of the ASCE in conjunction with the Geologan 97 Conference
352 pp, List, $35.00; ASCE member, **$26.25**
(#40255)

Observation and Modeling in Numerical Analysis and Model Tests in Dynamic Soil-Structure Interaction
● Toyoaki Nogami, Editor
Geotechnical Special Publication #64
Proceedings of sessions held in conjunction with Geo-Logan sponsored by The Geo-Institute of the ASCE
152 pp, List, $22.00; ASCE member, **$16.50**
(#40252)

Spatial Analysis in Soil Dynamics and Earthquake Engineering
● J. David Frost, Editor
Geotechnical Special Publication #67
Proceedings of sessions held in conjunction with Geo-Logan '97 sponsored by The Geo-Institute of the ASCE
144 pp, List, $40.00; ASCE member, **$30.00**
(#40258)

Unsaturated Soil Engineering Practice
● Sandra L. Houston and Delwyn G. Fredlund, Editors
Geotechnical Special Publication #68
Committee Report by the Subcommittee on Unsaturated Soils and the Committee on Shallow Foundations of The Geo-Institute of the ASCE; Proceedings of sessions on Unsaturated Soils
344 pp, List, $34.00; ASCE member, **$25.50**
(#40259)

ASCE
American Society of Civil Engineers

American Society of Civil Engineers 1801 Alexander Bell Drive, Reston, VA 20191-4400
Phone 800.548.2723 (ASCE), 703.295.6300; Fax 703.295.6333; email marketing@asce.org